CALIFORNIA
BURNING

CALIFORNIA BURNING

The Fall of Pacific Gas and Electric—

and What It Means for America's Power Grid

KATHERINE BLUNT

PORTFOLIO / PENGUIN

Portfolio / Penguin
An imprint of Penguin Random House LLC
penguinrandomhouse.com

Most Portfolio books are available at a discount when purchased in quantity for sales
promotions or corporate use. Special editions, which include personalized covers, excerpts,
and corporate imprints, can be created when purchased in large quantities. For more
information, please call (212) 572-2232 or e-mail specialmarkets@penguinrandomhouse.com.
Your local bookstore can also assist with discounted bulk purchases using the Penguin
Random House corporate Business-to-Business program. For assistance in locating
a participating retailer, e-mail B2B@penguinrandomhouse.com.

Library of Congress Cataloging-in-Publication Data

Names: Blunt, Katherine, author.
Title: California burning : the fall of Pacific Gas and Electric—and what it means
for America's power grid / Katherine Blunt.
Description: [New York] : Portfolio/Penguin, [2022] |
Includes bibliographical references and index.
Identifiers: LCCN 2022001979 (print) | LCCN 2022001980 (ebook) |
ISBN 9780593330654 (hardcover) | ISBN 9780593330661 (ebook)
Subjects: LCSH: Pacific Gas and Electric Company. | Wildfires—California. |
Electric power distribution—United States.
Classification: LCC HD9685.U7 P256 2022 (print) |
LCC HD9685.U7 (ebook) | DDC 333.793/20973—dc23/eng/20220119
LC record available at https://lccn.loc.gov/2022001979
LC ebook record available at https://lccn.loc.gov/2022001980

Printed in the United States of America
1st Printing

BOOK DESIGN BY CHRIS WELCH

For Russell and Becky,

whose generosity won't be forgotten

Contents

PART II

Cast of Characters

PG&E executives and employees

Brian Cherry: Vice president of regulatory relations, 2000–2014

Peter Darbee: CEO, 2005–2011; CFO, 1999–2004

Anthony F. Earley Jr.: CEO, 2011–2017

Bob Glynn: CEO, 1997–2004; various other positions

Bill Johnson: CEO, 2019–2020

Bill Manegold: Engineer, 1979–2014

Dan Richard: Senior vice president for public policy and government relations, 1997–2006

Nick Stavropoulos: Executive vice president of gas operations, 2011–2017; president and chief operating officer, 2017–2018

Geisha Williams: CEO, 2017–2019; executive vice president of electric operations, 2011–2017; senior vice president, energy delivery, 2007–2011

California regulators and politicians

Marybel Batjer: President of the California Public Utilities Commission, 2019–2021

London Breed: Mayor of San Francisco since 2018

Jerry Brown: Governor, 1975–1983 and 2011–2019

Paul Clanon: CPUC executive director, 2007–2014

Gray Davis: Governor, 1999–2003

Mike Florio: CPUC commissioner, 2011–2017

Sam Liccardo: Mayor of San Jose since 2015

Gavin Newsom: Governor since 2019

Steve Peace: Legislator, 1982–2002

Michael Peevey: CPUC president, 2002–2014

Michael Picker: CPUC president, 2014–2019

Arnold Schwarzenegger: Governor, 2003–2011

Prosecutors

Leading prosecution of PG&E on federal pipeline safety violations after the San Bruno explosion:

Hallie Hoffman: Criminal division chief for the Northern District of California

Jeff Schenk: Assistant US attorney for the Northern District of California

Hartley West: Former assistant US attorney for the Northern District of California

Leading prosecution of PG&E for its role in the Camp Fire:

Mike Ramsey: Butte County district attorney

Marc Noel: Butte County supervising deputy district attorney

Trial and defense attorneys

Steven Bauer: Corporate defense attorney who represented PG&E during the San Bruno trial

Kevin J. Orsini: Corporate litigation attorney who represented PG&E during its criminal probation following the San Bruno trial, as well as its second Chapter 11 proceeding

Mikal Watts: Personal injury lawyer who represented the largest number of wildfire victims during PG&E's second bankruptcy proceeding; led negotiations of the victims' settlement

Wildfire victims

Will Abrams: Vocal advocate for wildfire victims; lost his home during the Tubbs Fire

Kirk Trostle: Member of the Tort Claimants Committee during PG&E's second bankruptcy; lost his home during the Camp Fire

Early PG&E and Great Western Power Co. investors and advisers

James Black, 1891–1965: Great Western executive who led the company's merger with Pacific Gas and Electric Company and ascended to become PG&E's CEO

Eugene de Sabla Jr., 1865–1956: Established Pacific Gas and Electric Company following the 1905 merger of California Gas and Electric Corporation and San Francisco Gas & Electric Company

Guy Earl, 1861–1935: Attorney who helped build Great Western Power Company, rising to become its president

John Eshleman, 1876–1916: President of the California Railroad Commission; helped draft the Public Utilities Act of 1912, establishing regulatory oversight of California electricity providers

Julius Howells, 1859–1927: Hydroelectric pioneer who first envisioned a reservoir at Big Meadows and set in motion the effort to build Great Western's generation and transmission system

John Martin, 1859–1928: Established Pacific Gas and Electric Company following the 1905 merger of California Gas and Electric Corporation and San Francisco Gas & Electric Company

James D. Schuyler, 1848–1912: Hydroelectric expert who advised Howells on designing Great Western's power generation and transmission system

Judges

William Alsup: US district judge who inherited oversight of PG&E's criminal probation following its conviction on federal pipeline safety violations

Thelton Henderson: Former US district judge who sentenced PG&E to a five-year probation following its conviction on federal pipeline safety violations

Dennis Montali: US bankruptcy judge who oversaw both of PG&E's Chapter 11 reorganizations

Investors

Abrams Capital Management, Knighthead Capital Management, and Redwood Capital Management: Three hedge funds that together acquired a nearly 10 percent stake in PG&E during its second bankruptcy and pushed a restructuring plan that preserved the value of their equity

Elliott Management: Hedge fund that invested in PG&E's bonds during its second bankruptcy and competed with the company's shareholders to push an alternative plan of reorganization.

Centerbridge Partners: Investment firm that acquired equity in PG&E during its second bankruptcy as well as part of a line of credit Mikal Watts used to fund his wildfire litigation

Apollo Global Management: Private equity giant that invested in PG&E's bonds during its second bankruptcy, as well as part of a line of credit Mikal Watts used to fund his wildfire litigation

Other

Tim Belden: Former Enron trader who devised ways to manipulate California's energy market following its partial deregulation in 1998

Tom Dalzell: Former business manager of International Brotherhood of Electrical Workers Local 1245, PG&E's most prominent union

James Haggarty: Inspector with the San Mateo County District Attorney's Office; led federal investigation of PG&E's pipeline safety practices while working as a detective for the San Bruno Police Department

Tom Hoffman: Jury foreman during the San Bruno trial

David Hofmann: Organizational behavior expert at the University of North Carolina at Chapel Hill; advised the CPUC on issues related to PG&E's safety culture

George Sladoje: Helped build the California Power Exchange and became its chief executive during the energy crisis

Prologue

It's hard to say exactly when PG&E Corporation began its fall. Like the erosion of any great institution, it happened slowly, and then all at once as the weight of past mistakes became too much to bear. Few could have foreseen the extent of the consequences when California's largest and most powerful utility, a regulated monopoly with Gilded Age roots, caused a series of wildfires that killed more than a hundred people and razed hundreds of thousands of acres of vineyard and forest. The abject devastation revealed the company's systemic problems: chronic mismanagement, criminal neglect, existential risk. Its future had never looked so uncertain.

None felt that uncertainty more than the victims of the company's negligence. In California, utilities bear the cost of property damage and other liabilities if their power lines cause fires. The extent of the damage pushed PG&E to seek bankruptcy protection. It owed billions of dollars to those whose homes and families had been in the path of destruction, payments that would have helped them rebound. But hedge funds had swooped upon the company's distress, laying claim to most of the cash it had left. So PG&E decided to compensate victims with shares in the company itself.

Because of that decision, PG&E's former chief executive, an unflappable utility veteran named Bill Johnson, found himself face to face with a fire victim in a fluorescent-lit hearing room at the California Public Utilities Commission. It was February 25, 2020, an unusually warm winter day in San Francisco. The hearing was one of several the agency would hold as its staff considered the implications of PG&E's bankruptcy case, a federal proceeding that also required the involvement of state regulators. Johnson, who had been hired to lift PG&E out of the depths of crisis, appeared to field questions about the company's checkered safety record and its plan of reorganization, part of which involved funding a trust with equal parts cash and stock that would be slowly liquidated to pay fire victims over time. An hour of interrogation felt like three, the questions plodding and stoic. Then the victim, Will Abrams, spoke up.

Abrams, a management consultant with square glasses and small silver hoops in each ear, had spent months poring over thousands of pages of legalese, disappearing between Little League games and school functions to teach himself the language of PG&E. He was an interloper among the lawyers who had come prepared to interrogate Johnson about the technicalities of restructuring. When it was his turn, he signaled that his questions would be different.

Johnson, a former Penn State offensive lineman who, in his midsixties, still looked the part, hunched over the dais as Abrams explained why he cared enough to show up. PG&E had wronged thousands of Californians, few of whom had the inclination to battle a utility in the throes of bankruptcy. But Abrams had the uncommon resolve of a lone crusader. He had taken up the fight not just for himself, but for all who would soon hold shares in the company through the trust. He offered Johnson a metaphor: A man burns down someone's house and has only enough money to cover half the rebuilding costs. So he offers an investment opportunity, rife with risks, to cover the other half.

"Doesn't strike me as really fair," Abrams said. "Is that fair?"

Johnson responded carefully. "Fairness is often in the eye of the beholder," he said.

For Abrams, the irony was just too bitter. He had lost nearly everything late one October night in 2017, when a wildfire raced down a wooded canyon toward his neighborhood north of Santa Rosa, in California wine country. He awoke with his family in the early hours of the morning to a house billowing smoke. More than two years later, his young son was still in therapy for panic attacks, his family still in a rental house. PG&E's bankruptcy had kept them in limbo.

The fire that upended their lives was one of eighteen that ravaged Northern California that fall as gale-force winds and tree limbs battered PG&E's aging power lines, a consequence of branches growing too close to live wires. The lines sparked on contact, giving birth to infernos that killed twenty-two people. The winds returned the following fall to shake a power line in the foothills of the Sierra Nevada. Sparks flew as part of it broke. Within hours, the deadliest wildfire in California history had leveled the nearby town of Paradise. Eighty-four people burned to death.

When Johnson met Abrams in the hearing room in San Francisco, he had been CEO for ten months, enough time to grow accustomed to the hostility that permeated most any conversation about PG&E. He tried to defuse it with dispassion, staying calm and collected as he acknowledged the company's past failures and explained its plan to rectify them. He took the same approach with Abrams, who struggled to keep his emotion in check as he asked why victims should have to shoulder the risks that endangered the future of PG&E itself. The company's share price was sliding to a near-record low, having fallen more than 75 percent from its peak as investors realized that wildfires were not only hugely expensive but also potentially inevitable. Johnson was matter-of-fact. It was a complex situation, he said, made more so by the number of players involved. Attorneys. Investors. Politicians. Judges in two federal courts. Abrams sighed and prepared to pack up his notes.

"What can you do, as an executive, to ensure that these issues that victims are bringing up—that dramatically affect their lives— are at the forefront of how PG&E moves forward with a plan of reorganization?" he asked.

Johnson, whose soft voice offset his hulking figure, raised a finger, brow furrowed behind rounded glasses. He told Abrams he had visited Paradise five times after it was destroyed. Once, he said, he toured the area by bus with the company's board of directors, mostly New York financiers unfamiliar with the rural expanses of Northern California. During the ride, local officials played a recording of a 911 call from the Heffern family—grandmother, mother, and daughter. The women had huddled together in a bathtub and pleaded for help as the flames approached. They began to scream. Then the line went silent. The local sheriff showed a photo of their remains.

"If you think I am unaffected by this, by the victims and this treatment, you are wrong," Johnson said. "I'm going to do everything I can to make this right."

If it was an earnest promise, it was also a tired one. Each of Johnson's predecessors had pledged to do the same after a series of crises that had defined the past two decades for PG&E. None had succeeded.

Johnson was shepherding PG&E through its second bankruptcy in fifteen years. The reasons for the round trip were inextricably tangled. A series of executives had sought to please investors and politicians, often at the expense of customers. By the time the company confronted the risks of its aging electric grid, the problems were staggering. A protracted drought, exacerbated by climate change, had decimated its service territory, a heavily wooded region roughly the size of New England. One spark could set a whole forest on fire.

California lawmakers and regulators also failed to hold PG&E accountable, even as wildfire risk spread. They treated the company as a tool in their quest to preempt the long-term effects of climate change with ambitious renewable energy mandates. In doing so, they failed to recognize that a changing climate had made PG&E's power lines an immediate threat to the state.

PG&E's failure isn't just a California story. It is, in many ways, a harbinger of challenges to come as climate change exacerbates the vulnerability of the grid, built decades ago to serve a different era of electricity demand. Aging power lines across the West pose greater risks as drought-parched forests become more prone to catastrophic wildfires. And in the East and South, more destructive storms threaten to leave millions of people in the dark for days. In one way or another, all of the nation's investor-owned utilities are challenged to satisfy shareholders while making their infrastructure safer and more resilient. The very nature of the business creates tension between private interests and the public good, and PG&E is far from the only one that has struggled to strike the right balance.

Johnson, new to PG&E and to California, had been tasked with fixing what had broken, but he didn't answer to himself. The many people who had made a marionette of the company still held the strings as it worked to complete its most complex restructuring yet. Two months after meeting Abrams, Johnson announced he would step down as soon as the bankruptcy plan was approved. Abrams, meanwhile, had nowhere else to go.

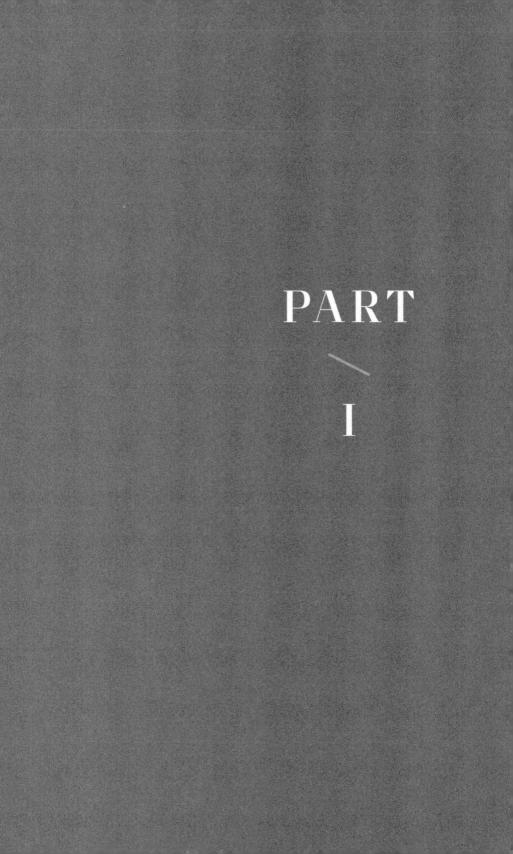

PART

I

INDICTMENT

brilliant flash broke the morning darkness on November 8, 2018, as strong winds pummeled a PG&E power line scaling the Sierra Nevada ninety miles north of Sacramento. A worn hook hanging from a century-old transmission tower broke clean, dropping a high-voltage wire that spit electricity just before sunrise. A shower of sparks set dry brush aflame.

PG&E recorded an outage on the line at 6:15 a.m. Five minutes later, one of the company's employees noticed a fire under the tower as he drove east on State Route 70, a remote two-lane road running through a steep river canyon that funnels mountain winds down to the valley below. Cell phone out of range, he radioed nearby colleagues, telling them to call 911.

The message reached the local fire captain at 6:29 a.m. He had risen early that morning, roused from sleep by the sound of pine needles pelting the roof of Fire Station 36. The two-engine outpost, tucked along State Route 70 at the mouth of the canyon, was clocking wind speeds as fast as fifty-two miles per hour. The call came as the captain was making breakfast for his small crew. It was already too late. Within fifteen minutes, both engines arrived across the river from the makings of a firestorm. There was no way to get ahead of it. The transmission tower, perched high along a steep, gravelly access route called Camp Creek Road, was almost completely inaccessible by fire engine. The crew struggled to stand upright as the winds whipped the flames. They were spreading with staggering speed. The captain radioed dispatchers for backup.

"This has the potential of a major incident," he said.

Within an hour, the fire had spread seven miles, burning through the small mountain communities of Pulga and Concow to arrive at the outskirts of Paradise, a tight-knit town of nearly twenty-seven thousand people nestled in the Sierra foothills. Residents awoke to emergency evacuation orders as softball-sized embers collided with dead trees. The fire was entirely out of control. At its fastest, it engulfed eighty football fields a minute, by some estimates. The tortured evacuation process began as thick black smoke took on a hellish orange hue. Escape routes became choke points, lines of cars inching along melting asphalt.

Dozens of people were left behind, unable to escape for reasons that made their gruesome deaths even more tragic. Many were in their seventies and eighties. One man had only just gotten his wheelchair out the front door. Another abandoned his wheelchair and tried to drag himself along the ground. A couple in their sunset years died together in their recliners, holding two dogs and two cats.

The fire overtook the town within hours. At noon, one of PG&E's first responders, called a troubleman, arrived at the ignition point in a helicopter. A regional supervisor had ordered emergency air patrol of the transmission tower in response to the early morning outage. The helicopter arrived to hover at a tall steel structure that, under normal circumstances, scarcely elicited anyone's attention. It had stood there for decades, becoming one with its surroundings, until the day it failed.

If you think about the grid as a network of roads, transmission lines are like highways, built to carry large amounts of electricity over long distances. They pick up electrons at power plants and channel them through thick, heavy wires held aloft by steel towers as tall as fifteen stories. The wires connect to substations, or off-ramps, where the power is reduced to lower voltages and distributed to homes and businesses through networks of smaller wires akin to local streets.

Transmission lines are subject to a cardinal rule. The wires must be kept away from one another as well as from the towers that support them. If the space in between them narrows too much, electricity can jump from wire to wire, or wire to tower, in what's known as an arc, a lightning-like bolt hot enough to melt metal and send sparks flying. To reduce that risk, the wires are suspended from strings of insulator discs hooked to the T-shaped arms of their towers.

Peering out of the helicopter, the troubleman saw an insulator string dangling. A hook about the width of a fist had broken nearly in half, dropping the insulator and the wire it held. An arc of electricity surged from the wire as it fell, scorching the tower in a blast of molten steel and aluminum.

The Camp Fire, named for the road near its place of origin, burned for seventeen days, destroying more than one hundred fifty thousand acres and nearly nineteen thousand structures, most of them homes. It didn't take long for investigators with the California Department of Forestry and Fire Protection to determine that PG&E's transmission line, known as the Caribou-Palermo, had almost certainly ignited it. Hours after Paradise had been consumed, a Cal Fire crew made its way to the tower where the PG&E helicopter had been seen hovering earlier that dark afternoon. Half a rusted hook dangled from the string of insulator discs still wired to the tower. The hook, which had hung from a hole in a long metal plate, was almost totally smooth at the point of fracture, evidence of a deep groove that had formed over decades. Millimeter by millimeter, the plate had cut into the curve of the hook, which was scarcely an inch in diameter. A jagged edge a few millimeters across showed just where it had broken.

The next day, the crew shared its findings with Mike Ramsey, Butte County's longstanding district attorney. Ramsey, a plainspoken prosecutor with stern white eyebrows, knew the county better than most anyone. His family had been there in the Sierra foothills for four generations to witness its evolution from a scattering of gold-mining settlements along the Feather River to a bucolic spread of communities home to more than 226,000 people. Ramsey had grown up exploring the forests with his father, the local game warden, learning the space where civilization dissolved into wilderness. The younger Ramsey had left Butte County only briefly, for college and law school, returning shortly after graduating when a position opened in the district attorney's office. He assumed the post in 1987 and had been reelected ever since. Upon taking office, he adopted a mission statement: "To do justice, as no one is above the law, nor beneath its protection."

Like other counties atop California's gold fields, Butte County owed much of its early development to PG&E. The company's founders had spent the last decade of the nineteenth century exploring the canals and waterwheels that supported the region's gold miners and buying up the companies that operated them. One by one, they built out a series of hydroelectric powerhouses that generated electricity for Oroville and Chico, Butte County's largest cities, as well as neighboring counties whose populations had exploded during the gold rush. That network formed the foundation of a sprawling monopoly that would grow to encompass almost all of Northern and central California. With it came the goodwill of millions of customers as PG&E expanded its services, electrifying industries and creating jobs across the state for thousands of proud engineers.

Ramsey had watched that goodwill dissipate in the years after he became district attorney. It started in the 1990s, when PG&E power lines caused several destructive wildfires in the Sierra foothills north of Sacramento. Around the same time, a young law clerk named Erin Brockovich discovered that PG&E had allowed water tainted with hexavalent chromium, a carcinogenic metal, to leach into a small town's groundwater supply in the Mojave Desert outside of Los Angeles. Brockovich helped residents sue PG&E, a story that became a Hollywood blockbuster starring Julia Roberts. In 2010, a decade after the movie's release, one of PG&E's natural gas pipelines exploded in San Bruno, south of San Francisco, destroying a neighborhood and killing eight people. Then, in 2017, more than a dozen deadly fires swept Northern California after wind-whipped branches collided with PG&E power lines. One of them started just outside of Paradise when an oak branch fell on a transmission line. Ramsey spent the summer of 2018 negotiating a settlement with the company, which might have broken state law in failing to clear the branch. It agreed to fund a power line inspection program headed by the county fire department.

Ramsey recognized that the Camp Fire was different than the vegetation fires of years past. It seemed to him that a company with a history of flouting the rules might be hiding yet another set of problems. Ramsey told the Cal Fire team that his office wanted to join in their investigation. He told them to treat the transmission tower as a crime scene and to prevent anyone, including PG&E employees, from entering unaccompanied. The next week, Cal Fire supervised as company line workers began dismantling the tower. The investigators seized hooks and hanger plates as evidence.

Ramsey's right-hand man was Marc Noel, who had started working for Ramsey just a few years after he took the top job. Noel had grown up in the Bay Area but made Butte County his home in 1991 and became deputy district attorney shortly thereafter. Nearly thirty years in the role had given him a gray goatee and a habit of spitting tobacco into empty cups and cans. He truly loved Butte County, which was sewn of a wholly different fabric than that of San Francisco. It was equal parts red and blue, with farms and fruit orchards all along the valley floor. It was not a wealthy, bustling enclave. The allure was its simplicity: kind, honest people living in a beautiful and secluded corner of the state.

Noel had long suspected that PG&E didn't understand or respect that about Butte County, or any other part of the rural foothills. He had gotten a feel for the way the company operated after various vegetation fires, which resulted in slick lawyers heading north from San Francisco or Los Angeles to show up the local prosecutors. Not long after the Camp Fire, he picked up Malcolm Gladwell's *David and Goliath*.

Over the years, Noel had become something of a fire expert. He had been working arson cases since the 1990s after training with Cal Fire to investigate cause and spread. Transmission lines, though, were another matter. Never had he thought much about the hundreds of steel structures carrying high-voltage wires through the forests. He began reading anything he could find, painfully aware of how much he had to learn.

In the weeks after the fire, Ramsey and Noel brought in experts from the FBI to assist in analyzing evidence. Noel took the hook and other parts collected from the suspect tower and showed them to an FBI metallurgist. It was obvious how the hook had failed. The question was whether PG&E could have prevented it from happening. The metallurgist told Noel his team would need to collect parts from towers in the vicinity. Without a basis for comparison, it would be impossible to tell whether the fire was the result of the company's negligence or simply a tragic accident.

So Noel and Ramsey devised what they called the Exemplar Tower Project, a sweeping search for similar towers along the canyon. Noel boarded a county helicopter on New Year's Eve and instructed the pilot to fly low and slow along the Caribou-Palermo. He had learned, by that point, that hooks like the one that broke are supposed to fit snugly in the holes of their hanger plates. As he flew, Noel noticed something suspicious. The holes looked especially large. It was a sign that the hooks had been hanging for decades, wearing down little by little with every windstorm. Three towers in particular seemed especially similar to the one where the fire had started, with gaps between hook and plate large enough to stick a finger through. Ramsey and Noel planned to seize parts of them as evidence.

PG&E launched its own investigation as Ramsey and Noel began theirs. In December, not long after the fire stopped smoldering, workers set out to inspect the Caribou-Palermo in its entirety. They climbed transmission towers perched high on the rocky, forested slopes of the canyon to look closely at the tiny pieces of hardware holding the wires aloft. The diagnosis was devastating. The linemen discovered more than a dozen critical hazards, mostly involving hooks and other connectors that had been in place for decades.

Across the country, transmission lines are among the oldest parts of the grid. Many were constructed in the years after World War II as Americans moved from cities to suburbs, built homes wired with wall sockets, and bought new electric appliances. Some transmission lines are even older, developed shortly after the turn of the twentieth century to replace gas lamps and candles at a time when electricity was still something of an experiment.

The Caribou-Palermo, a fifty-six-mile conduit running along the rugged edge of the mountain canyon, was a relic of that era, so old that it was once considered for the National Register of Historic Places. It had been built around 1921 by a company called Great Western Power, which had competed with PG&E until the two companies merged in 1930. The line was part of a transmission network carrying electricity from hydroelectric powerhouses along the Feather River, which cascades down the slopes of the Sierras, losing thousands of feet in elevation in a steep, rapid plunge. The line picked up electricity at the Caribou powerhouse, where three turbines harnessed the churn of the river, and carried it south to San Francisco to serve a booming population.

PG&E shut the Caribou-Palermo down upon discovering the extent of its problems. Employees began a frantic search for construction and maintenance records. They came up short. The Caribou-Palermo files were incomplete. And it wasn't just that line. The company also lacked records on dozens of others.

PG&E is hardly the only American utility with poor records. It's the arthritis that plagues the companies that deliver electricity and natural gas to millions of people. Many of them have been painfully slow to organize and digitize the millions of pieces of paper they have retained over the course of decades. For those whose networks of wires and pipelines span tens of thousands of square miles, it's a gargantuan, costly task that only grows more difficult with each passing year. Perhaps their employees, siloed in their various divisions, use different systems to search and file, raising the prospect of a nightmare integration process. Maybe they unwittingly disposed of critical records years ago, creating hard-to-fill information gaps. Or maybe the trouble lies not with the records themselves, but the company's means of creating them. Records are only as good as the inspection and maintenance work they are made to reflect.

In any case, there's a common reason for the inertia. Utilities have little incentive to overhaul operations that have worked reasonably well for years. As monopolies, they are not subject to the sort of relentless competition that drives unregulated companies to improve or innovate. Poor records, whether the result of information loss or shoddy inspection work, don't often result in much more than a power outage here, a gas leak there. It takes a catastrophe to reveal the extent of the problem.

Such events have forced a number of utilities into high gear in recent years. In 2015, tree branches brushed against a distribution line owned by a small rural utility in north-central Washington,

igniting the 11,200-acre Twisp River Fire that killed three firefighters with the US Forest Service. Prosecutors alleged the utility hadn't
done enough to record the threat and address it. That same year, a
well in an underground storage tank owned by the nation's largest
gas utility began to leak about thirty miles north of downtown Los
Angeles, releasing invisible plumes of methane, a potent greenhouse
gas that forced hundreds of nearby residents to evacuate. The leak
occurred because the casing of the well, drilled in 1954, had ruptured
as a result of severe corrosion, something the utility's inspectors had
never examined. Another disaster occurred in 2018, when a series of
destructive explosions rocked the Massachusetts towns of Lawrence
and Andover after utility workers replacing a century-old gas pipeline
made a critical error. The engineer overseeing the project had relied
on a digital database that turned out to have holes—the records
needed to safely complete the project were held as hard copies elsewhere.

Such tragedies have attracted more attention as regulators and
politicians grapple with how to deal with the nation's aging infrastructure as climate change further stresses a system already vulnerable to failure. The Twisp River Fire was part of a larger wildfire
complex that blazed fast and hot through parched, overgrown forests
struggling through an unusually dry summer. The Southern California gas leak was the worst in US history, releasing the emissions
equivalent of more than a half-million cars driving around for a year.
And the Massachusetts explosions exposed the risks that old pipelines continue to pose even as the nation reduces its reliance on
natural gas, raising questions about who should continue to pay for
system maintenance.

No utility, however, has ever faced a disaster as deadly and ominous
as the Camp Fire. The results of the Caribou-Palermo inspection sent

a wave of panic through PG&E's top brass. Company managers had been trying for years to improve record keeping by uploading thousands of documents into centralized databases. But the work had been slow and was hampered by the shortcomings in the company's inspection processes. The problems discovered on the Caribou-Palermo were a sign of how short those efforts had fallen.

PG&E soon announced plans for a massive inspection blitz covering every part of its electric system in areas at high risk of wildfire. A decade earlier, those areas had been confined to a few thickets within the forests that blanket much of Northern California. Then a severe drought set in, turning trees into standing firewood vulnerable to the smallest of sparks. The devastation encompassed most of the coastal and mountain regions surrounding the Central Valley, putting more than half of PG&E's 70,000-square-mile service territory at risk. PG&E had 5,500 miles of transmission lines traversing the riskiest areas. Its distribution network spanned far wider. The company promised to finish the inspections before the early summer start to California wildfire season. That meant completing years' worth of work in a matter of months. Workers fanned out in the final weeks of 2018.

In June, the company released an apologetic statement. It told the public that the inspections had revealed the need for more than 250,000 repairs across the system. Ten towers supporting a transmission line near the Golden Gate Bridge needed complete replacement. And the Caribou-Palermo, riddled with problems, would never run again.

In the spring of 2019, as PG&E started its inspection blitz, Ramsey and Noel had seen enough transmission towers to conclude that the one where the fire started wasn't alone in its disrepair. It was time to start collecting parts as evidence. They returned to one of the three towers that had appeared especially run-down during their helicopter flights. But the old hooks and plates were nowhere to be seen. PG&E had replaced them with new parts, even though it had deemed the line too unsafe to operate.

Noel was infuriated. When it came to PG&E, it didn't take much to set him off. Any hint of obstruction sent him reaching for another tobacco plug. He dialed up PG&E's attorneys to ask why the company had taken parts from the tower. PG&E was preserving them for use as evidence in other litigation, one of the attorneys replied. If Noel wanted to take a look, he would have to drive several hours to a warehouse where PG&E had stored them. Noel felt his blood pressure rising. He demanded return of the evidence. The attorney shifted to sarcasm. "Where would you like us to deliver it to you, Marc?" he asked.

"I'd like you to deliver it to the FBI national laboratory in Quantico, Virginia," Noel replied.

The line fell silent for several seconds. "Are you serious?" the attorney finally asked.

"Yeah," Noel replied. "Fucking serious."

The company's attorneys changed their tone. They had been trying to determine the extent of the prosecutors' probe, and the FBI's involvement came as an unwelcome surprise. The attorneys allowed the prosecutors to send for the evidence. Within forty-eight hours, it was ready to be shipped off to Quantico. Noel smiled to himself, daring PG&E to underestimate him again.

At the end of March, Ramsey empaneled a special grand jury to determine whether the company's neglect amounted to criminal conduct. Of one hundred potential jurors, Ramsey narrowed it to nineteen.

They had a year to deliberate. They immediately subpoenaed PG&E for all of the records it possessed on the origin and maintenance of the Caribou-Palermo. PG&E buried them in information, sending over enough documents to fill several tractor trailers. It overwhelmed the computer storage in the district attorney's office. Ramsey had to buy special servers to hold it all.

Much of it wasn't relevant to the investigation. The way Noel saw it, PG&E had put the needles in not one haystack, but every haystack in the field. And some needles were missing altogether. After sifting through millions of documents, the prosecutors could draw but one conclusion: PG&E had almost no records on the age of the hook that failed. The company couldn't say for sure just how long it had hung there, rocking on its plate in the windy Feather River Canyon.

At Quantico, the FBI analyzed the makeup of the hook. The results revealed that it was nearly as old as PG&E itself. It had been manufactured by the Ohio Brass Company, which started in 1888 as a small foundry forging buggy harnesses and plumbing valves before shifting to hardware for some of the first transmission lines ever built. Historical records indicate that the hook had been purchased for 56 cents in 1919. By the time it finally gave out, it had weathered nearly a century's worth of wind and precipitation. The hook was no anomaly. PG&E found no records of ever replacing parts on the tower where the Camp Fire started, save for pieces of hardware meant to connect parallel wires.

The company hadn't documented the extent of the problem in part because its employees hadn't seen it. PG&E couldn't locate records on Caribou-Palermo inspections prior to 2001. Since then, workers had periodically inspected the line from the ground and from the air—but not by climbing the towers, one of the best ways to examine the small hardware supporting the wires. PG&E reserved that treatment for transmission lines showing serious problems. By that standard, the Caribou-Palermo should have warranted it. As early as 2007, a PG&E engineer sought to replace nine of its towers standing along a remote mountain road that zigzags off the Feather River Highway. In a plea for funding, the engineer noted that "failure is imminent due to the age of both the towers and the conductor."

"We could be picking up these towers out of the Feather River Canyon when they fall over," the engineer wrote in an internal email. PG&E nixed the effort, saying the transmission division had "reprioritized" its slate of projects.

In 2012, a strong winter storm caused five of the towers on the line to topple over in a tangle of wire and steel. They weren't the ones earlier slated for replacement, but they were just as old, located a few miles away from the Camp Fire ignition point. The PG&E engineer who investigated the mess didn't write a formal report on what had happened. Instead, he advised colleagues in an email to inspect the other towers for defects. They conducted no such inspections.

The Caribou-Palermo showed more problems, year in and year out. In June 2018, one of the company's transmission supervisors finally called for inspectors to climb each of its towers. PG&E couldn't find records on why the supervisor ordered the work, saying only that the age of the towers factored into the decision. By November, inspectors had worked their way through eighty of them along the edge of the mountains south of Paradise. They would have continued to the ones lining the Feather River Canyon. But nature didn't wait. The work came to a cruel halt on November 8.

The evidence proved that PG&E had neglected the line even as it deteriorated to the point of danger. For the grand jury, the question was whether it did so recklessly—the legal standard that would determine whether the company should face homicide charges. The jurors were unequivocal. They concluded that PG&E was well aware that its negligence had created a serious fire risk in the Feather River Canyon but did almost nothing to mitigate it. On March 17, 2020, the grand jury indicted PG&E on eighty-four counts of involuntary manslaughter. The company agreed to plead guilty.

Ramsey offered a pithy summary in his investigation report. "In 1930, PG&E blindly bought a used car," he wrote. "PG&E drove that car until it fell apart."

Two

BUILDING A
MONOPOLY

I n 1882, a young engineer named Julius Howells joined a band of
explorers on a trek through the Northern Sierra. They had set out
at the behest of a Harvard geology professor fascinated by the hy-
drology of mountains, forged from volcanic activity millions of years
earlier. As Howells traveled along the Feather River, he came upon a
lush valley nestled high near the peak of Mount Lassen. The river
basin, called Big Meadows, collected water from mountain streams
and underground springs erupting from the porous volcanic terrain.
A fifty-square-mile expanse of fields and cattle ranches formed the
base of a five-hundred-square-mile watershed. Howells thought not
of its academic importance, but of its commercial potential.

Howells, a Hoosier who had made his way to California by helping railroad companies site their tracks across the country, was fascinated by hydropower. He saw enormous opportunity in the Northern
Sierra, where tumbling rivers lost thousands of feet in elevation in

their rapid downhill journeys. The region's miners had spent several decades diverting the cascades to feed enormous hoses used to blast gravel from gold deposits. Their flumes and canals formed the skeletal infrastructure for the state's first hydroelectric power systems.

For tens of thousands of years, running water has given rise to some form of industry. The simple mechanical design of the waterwheel freed men and animals from the manual labor needed to irrigate crops and mill grain, lumber, and textiles. Century by century, inventors honed the wheel's efficiency. By the early 1800s, most every mill and factory drew power from large wooden wheels positioned to catch fast-moving water. Then, a series of engineers modified the wheel to function underwater as a turbine, a rotary engine with large metal blades. Situated at the base of water chutes or tunnels, the turbines produced huge amounts of mechanical power.

The design evolved during the heyday of electrical innovation. Men around the world were experimenting with the dynamo, the precursor to the modern generator. Each iteration of the device proved more effective at transforming mechanical energy, most often supplied by a coal-fired steam turbine, into electrical energy, setting the stage for Thomas Edison's pursuit of commercial lighting systems and power plants. It wasn't long before someone recognized that an underwater turbine, like a steam turbine, was perfectly suited to power a dynamo. Hydropower would become hydroelectric power.

In 1880, Edison founded the Edison Illuminating Company to equip New York City with incandescent light. Within two years, he had begun stringing the wires for the world's first centralized power plant on Pearl Street in lower Manhattan. His followers, some of whom had bought the patent rights to install Edison equipment in other major cities, watched with interest. Coal-fired steam engines would power the dynamos, which would light several hundred bulbs for some eighty-five customers in the city's financial district.

Henry James Rogers, the president of a local paper company in Appleton, Wisconsin, first heard about Edison's plans for Pearl Street Station while bass fishing with an old friend who worked for the Western Edison Light Company in Chicago. Rogers, who also controlled Appleton's gasworks, had been waiting for the day when electric power could be deployed at scale. A few years earlier, he had been hesitant to invest in the gas plant that lit the town's street lamps, knowing that electricity might soon render it obsolete. As soon as he returned from his trip, Rogers purchased the Edison patent rights to light Appleton and other towns along the Fox River. He started with a single Edison dynamo to light one of his paper mills as well as his newly built mansion.

In August 1882, Rogers and a group of local investors contracted with Western Edison to wire the dynamo to a waterwheel powering equipment at the paper mill. It was a risky experiment. Rogers left the task to a Western Edison crew and steered clear of the mill that day out of concern about what might happen. The system worked, but in fits and starts. The mill equipment interfered with the circuitry and caused the lights to dim and surge. Dozens of new, expensive bulbs burned out.

The next month, Rogers had the crew move the dynamo into an old office shed on the mill property. The men hooked it up to its own waterwheel to stabilize the current and ran strands of bare copper wire to the mill and Rogers's home. The buildings lit up at the end of September, just weeks after Edison's Pearl Street Station was complete. It was the nation's first hydroelectric power plant. A reporter from the *Appleton Crescent* visited the site a week later to find Rogers and his investors thrilled with their setup. "The electric light may entirely supersede the use of gas as an illuminator in our city," the reporter wrote. "But that remains to be seen."

For that to happen, Appleton would need better wires. Edison and others had for years been grappling with how to transmit power over long distances. Solving that equation would allow for power plants to be built in remote areas rather than crowded cities. But Edison had developed his electric lighting systems to run on direct current—low-voltage electricity moving in a single direction. It was difficult to bump up to higher voltages necessary for successful transmission. Nikola Tesla, a Serbian immigrant who had briefly worked for Edison, was a step ahead of his former employer. He championed alternating current, which periodically reverses directions and changes voltages more easily. Tesla licensed his technology to George Westinghouse, who pushed to outdo Edison in the rush to solve the transmission puzzle. The battle ended in 1893, when Westinghouse won the contract to build a hydroelectric plant at Niagara Falls with a twenty-two-mile transmission line to Buffalo, New York.

As the plant got underway, another pioneering development was taking shape in California. Hydroelectric power held particular potential along the West Coast, which had almost no coal deposits to power steam turbines. What the region lacked in hydrocarbons, it made up for in hydropower. It would become known as "white coal."

In 1895, the Sacramento Electric, Gas and Railway Company opened California's first powerhouse on the banks of the American River in Folsom, northeast of Sacramento. The company had built a twenty-two-mile transmission line to carry the electricity to the state capital. The powerhouse, which came to life several months before the one at Niagara, celebrated its opening with an "electric carnival" capped with a parade of illuminated floats. The *Sacramento Record-Union* called the celebration "indescribably brilliant," musing that the "meaning of it all reaches far away into the future, the dawn of which gilds the housetops of Sacramento already, and foretells the rising sun of prosperity."

Howells surveyed the Sierra for two years after the other explorers disbanded, awed by the sheer bounty of water. He endeared himself to the small town of Quincy, high in the Plumas National Forest, by playing the fiddle at social dances. But he soon left California for a civil engineering job in his hometown of Richmond, Indiana. He moved to Chicago, and then New Mexico when a position opened at the Santa Fe Water Company. He became its chief engineer, helping to design some of the company's first dams. He did the same for the San Diego Flume Co. and the San Joaquin Electric Company in Northern California. In 1901, he made his way back to Big Meadows to study and measure the watershed. He returned to his office in Los Angeles with plans to flood the river basin. The enormous reservoir would be his most ambitious project yet.

It was the sort of undertaking that would require a blank check and the right connections. Howells would have to bring more engineers on board. He would need to sell investors on his vision. And he would have to secure the rights to use the land and water at Big Meadows.

Howells got in touch with James Schuyler, a hydraulic engineer who had just written a book on storage reservoirs. Schuyler was best known in the American West for building dams and waterworks throughout the region, but he had also consulted abroad on major projects in Mexico, Brazil, and Japan. When it came to water, he was an explorer and a romantic, often toting a camera on work trips. In Japan, he fell seriously ill while consulting on a reservoir high in the mountains and had to be rushed to the city of Kobe in a flimsy skiff. He spent the entire woozy journey snapping photos of the tumultuous river.

Schuyler considered Howells's proposal and agreed it had huge potential. It would also be hugely expensive. Schuyler knew someone with deep pockets and an appetite for risk: a Southern California business magnate who had made a fortune from a fruit-shipping company he had founded some years earlier. Schuyler scheduled for him to meet Howells, who succeeded in impressing his potential investor with the scope of the project. The dam, Howells explained, would release water to downstream powerhouses that would churn out endless amounts of electricity. A transmission network would carry that power two hundred miles to San Francisco, where demand was poised to skyrocket. Schuyler called it a "tempting field for electric transmission enterprises to exploit."

The business magnate then called on his brother, Guy Earl, an attorney in Oakland and man about town who bore some resemblance to the actor J. K. Simmons. Earl, who served on the board of regents of the University of California, Berkeley, was politically active with social connections all throughout the Bay Area. He wore many hats. Once, he had to figure out what to do with a dead African bull elephant that Theodore Roosevelt had decided to give to Berkeley. ("Fear a hunter bringing gifts!" Roosevelt wrote in a letter to the regents.)

Earl wanted in on Howells's project. It was such a good idea that it would have to be kept a secret to guard against competition, and

Earl would be one of the few in the know. He happened to be friendly with some prominent stockbrokers whose firm had offices in San Francisco and New York. He quietly tapped them to help shore up financing.

In a letter in March of that year, Howells assured his backers that the venture would be hugely profitable. "This will by far be the largest and boldest water-power development yet proposed in California, and, as far as I know, when built, will stand second only to Niagara," he wrote. "Taken into consideration the enormous amount of power proposed, the simplicity of the physical arrangement, and the cheapness of its various parts, together with its reliability in case of years of excessive drought, make it stand in a class by itself."

Earl enlisted an Oakland real estate investor to purchase the land for the reservoir. The first to sell was Augustus Bidwell, whose family owned a large part of Big Meadows. Bidwell persuaded other landowners to sell and agreed to help secure the water rights. At the time, the requirements for doing so required more luck than formality— one simply needed to tack notice of his claim on a tree near the point where he intended to divert the water, and then inform local officials of when and where it had been posted. Prior in time, prior in right, as the saying went.

In the spring of 1902, Howells and Bidwell set off to find a tree, traveling by sled through deep mountain snow. Howells was hardly expecting competition. His plans for the reservoir had been closely guarded. But coincidence prevailed. Another pair arrived in the vicinity on the same day, and Howells and Bidwell soon realized they weren't alone in their pursuit. They posted their water claim in Earl's name. Two miles downstream, the strangers posted their own notice. Howells set off to post another claim nearby while Bidwell raced their competitors to Quincy, the Plumas County seat, trailing their sled for thirty-two miles in a race worth millions. Bidwell only just eked out a victory.

That summer, Schuyler presented Earl with a forty-three-page report charting careful rainfall calculations and measurements of the Big Meadows watershed. He had surveyed the area and compiled a spread of black-and-white photos of the valley that would become the storage reservoir, with its thick forests and snow-capped mountains. He included a shot of the straggly tree where Howells and Bidwell had tacked their water claim, its trunk displaying the small scroll of paper that set the project in motion.

Schuyler pegged the whole project at $21 million—a fortune by the standards of the time. The powerhouses, equipment, and labor would each cost hundreds of thousands of dollars. The generators and transmission lines alone would cost several million. But Schuyler believed it would be worth it.

"The more carefully I have considered the general plans of power development by the Western Power Co., the more deeply impressed I am with their simplicity, their comparative cheapness per unit of power, and their entire feasibility," he wrote. "Thus far, the demand for electric power has surpassed the utmost capacity of all the plant projected for its supply, and the future outlook for marketing the entire output of this Company seems quite encouraging without the indulgence of excessive optimism."

In 1904, the local stockbrokers spoke with Earl and used their New York connections to secure a large investment from a railroad magnate who sat on the board of the Western Pacific Railroad, which was building a new route that would run from Oakland to Salt Lake City through the Feather River Canyon. The railroad magnate, who understood the region's massive development potential, became the power company's president and largest shareholder. A group of prominent

New York financiers soon lined up to back the project. They included executives at some of the largest concerns of the day—the American Tobacco Company, Standard Oil, and Comstock mines. With that, the Western Power Company was born. Two years later, it was re-christened the Great Western Power Company with $25 million in capital.

The men of Great Western had competitors. John Martin and Eugene de Sabla Jr., an unlikely yet formidable pair, had spent years laying the foundation for what would become Pacific Gas and Electric Company, building their own electricity network and snapping up tiny, local power companies in a quest to dominate Northern California.

De Sabla came from money. His wealthy relatives had fled from France to the West Indies and wound up owning plantations in South and Central America. Born in Panama, de Sabla moved to San Francisco in 1870 at age five with his father, who started a coffee import business. His father enrolled him in the city's best schools, thinking he was destined for the Ivy League. But de Sabla had little inclination toward college, despite being smart and fiercely competitive. He was more interested in mining. Tall and brawny, he started off working long hours at an Arizona copper mine. He shoveled coal to power mine equipment, sometimes for as little as a dollar a day, until the financial panic of 1893 sent copper prices plummeting and forced the mine to close.

At twenty-one, he returned to San Francisco to liquidate his father's coffee business, which had sunk into debt. He needed a new venture. He found an opportunity through his former boss at the copper mine, who partly owned another mine on the South Yuba River in the Sierra foothills. Underground springs made the mine difficult to operate. The two men decided to build a hydroelectric plant to pump the water out and supply power to nearby towns.

Martin, meanwhile, was born in the Midwest and raised by an aunt in Brooklyn after his parents died. He was scrappy from the start. At thirteen, he took off on his own, selling newspapers on the street to support himself. He traveled to Alabama and then Illinois, working jobs in real estate and meatpacking. He settled in California in 1891 and established an iron pipe business. He loved engineering and had a knack for learning despite having no formal education.

A mutual friend introduced Martin and de Sabla over lunch in San Francisco in 1894, thinking de Sabla might need to order pipes for the hydroelectric power project. Martin, a sturdy man with wide brown eyes and a thick, dark mustache, arrived a half hour late—he had stopped to witness the birth of his fifth child. The meeting was otherwise unremarkable. De Sabla described the powerhouse proposal, mentioning that he hoped to obtain the electrical equipment from the Stanley Electrical Manufacturing Company, which he considered best in class. Martin, who was supposed to have used the time to pitch his pipe business, was instead more eager to ask questions about electricity. The men parted ways with no pipe order placed and no plans to meet again.

Months later, Martin and de Sabla met by chance on the street in San Francisco's financial district. "I'm ready now for your order," Martin said, taking de Sabla by surprise. As it turned out, Martin had spent weeks crafting a plan to get involved in developing the powerhouse. He had traveled to the Stanley Electrical headquarters in Massachusetts and returned with the rights to sell its products in California. Martin was prepared to supply not just iron pipe but all of the electrical equipment necessary for the project. De Sabla gave Martin the contract.

Neither man knew what he was doing. Martin called upon colleagues at Stanley Electrical for a crash course while de Sabla solicited customers. They somehow made it work. In 1896, the powerhouse began serving the nearby towns of Nevada City and Grass Valley. Martin and de Sabla immediately set to work on building another powerhouse just a few miles east along the same river. Martin hired three young men to construct it and sent them into the mountains, where each was given four donkeys to help carry cement and lumber up the slopes. It was round-the-clock work. They lived on site, growing their own vegetables. Every now and then, a salmon tacked to a wooden plank would float down the river to the powerhouse, a gift from dam workers upstream. The little town of Marysville, just outside of Yuba City, lit up when the men installed the first generator.

As different as they were, de Sabla and Martin were both utterly tireless. Electricity became their shared obsession. They fed on each other's energy, building new powerhouses and buying up competitors. In 1901, they made inroads into the Bay Area by building a 142-mile transmission line to carry power from the Sierra foothills to Oakland. The line crossed over water at the Carquinez Strait, part of the estuary that feeds the San Pablo Bay north of San Francisco. To clear the channel, Martin and de Sabla had tall steel towers built on each side and commissioned barges to drag the cables from shore to shore. It took five days to unspool the wires and hoist them into position with enough room for ships to pass beneath. The wires spanned roughly 4,400 feet from tower to tower, an unprecedented distance.

By the end of 1901, Martin and de Sabla formed the California Gas and Electric Corporation, which consolidated all the smaller power companies serving customers in the Sierra foothills and parts of the Bay Area. They had their sights on bigger markets. For months, Martin had been eyeing the Oakland Gas Light and Heat Company, which served retail markets throughout the East Bay. Acquiring it would allow California Gas and Electric to close in on San Francisco.

The city had a longstanding incumbent: the San Francisco Gas & Electric Company, which had gotten its start decades earlier by supplying the gas to light some of the nation's first streetlamps. Once the city's dominant power supplier, San Francisco Gas & Electric had been buffeted by competition. Its fleet of steam-operated power plants couldn't compete with the lower-cost hydroelectric power that California Gas and Electric produced in abundance. The two companies merged in October 1905, creating Pacific Gas and Electric Company.

Then tragedy struck. On April 18, 1906, the San Andreas Fault shifted along the California coast. A 7.9-magnitude earthquake shook San Francisco with devastating force. It was a shock unlike any other. Power lines fell and ignited fires. Substations were reduced to wire and rubble. Gas pipelines ruptured and exploded. All but two of PG&E's power plants were destroyed. It was a major setback for the newly expanded company, which had just issued debt to complete the merger. Some within the company thought that bankruptcy was the only option. But Martin and de Sabla decided instead to tap shareholders for more funds and then take on more debt. They would pay it back with revenue from a city rebuilt.

At first, the destruction in San Francisco spooked Great Western's investors. If the city had been all but leveled, where would the company sell its hydroelectric power? But Guy Earl, then the company's vice president, convinced them of their fortuitous position. In a lengthy handwritten letter, he explained that PG&E had been brought to its knees, its San Francisco stronghold weakened. By the time it managed to rebuild, Great Western would be at the gate. By the fall of 1906, the company had started building its first powerhouse at Big Bend, eighteen miles up the Feather River from Oroville. It was the first piece of what would become a massive system supported by the storage reservoir at Big Meadows.

Howells had known from the beginning that competition would be fierce. Before work began, he had written to Earl to inform him that businesses in San Francisco paid between fifty and ninety dollars per year for a horsepower of electricity, a measure equivalent to 746 watts. That reflected the cost of coal or fuel oil needed to power steam turbines. Because hydropower was essentially free, Great Western could produce electricity more cheaply, Howells wrote, "making it possible, to sell it, in the case of a rate war, as low as $15 per horsepower per annum, at a profit."

Great Western would also have to win over customers not just in the city, but throughout the Northern California towns that PG&E served. In 1908, as Great Western was completing the Big Bend powerhouse, the company invited residents of Oroville and Marysville to come and see it. Western Pacific had by then completed its railroad along the Feather River Canyon. More than 150 people boarded a train and rode it into the mountains. Great Western executives allowed them to walk inside a massive tunnel that would channel water from the Feather River to the turbines. On the way back, the train stopped so that the passengers could hop off and pick berries growing alongside the track.

"Those who made the trip are loud in their praise of the treatment accorded them by both the officials of the Western Pacific and Great Western Power Company," the *Oroville Daily Register* reported. "Despite the fact that these same people had read time and time again of the operations of the companies, they were astonished at the work that they saw."

Great Western had also been busy finding crews to build a two-hundred-mile transmission line, even tapping students from Stanford and Berkeley to help to erect the giant steel towers. The line snaked like a great artery from the Sierra foothills to Oakland, passing through Sacramento, Oroville, and Marysville on the way. One of the segments was built inward from each end, connecting in the middle above the tiny hamlet of Nicolaus. The oldest villagers watched in awe one summer day as crews hoisted the wires, a sign of big changes to come.

At first, Great Western and PG&E existed amicably. PG&E bought some of the newcomer's wholesale power and distributed it to retail customers. But Great Western quickly encroached upon PG&E's territory. In 1911, it swept into Sacramento and underbid the older company to power the California State Capitol. It would soon target Marysville, Oroville, and a scattering of other cities throughout the North Bay. It was prepared to take on San Francisco, but it had no retail outlet. So, like PG&E had done, it secured its position by acquiring one of the remaining companies there: City Light & Power, founded in 1907 after the earthquake created room for competition.

"The Great Western now constitutes the most formidable competitor the Pacific Gas and Electric has ever had to face," *The Sacramento Bee* declared when news of the acquisition broke.

That year, word got out that PG&E had approached Great Western about a possible merger to form what would have been the largest power company in the world at the time. The idea made sense. PG&E controlled much of the electric distribution system throughout Northern California, while Great Western was developing one of the most valuable hydroelectric networks anywhere. In September, Great Western's president hosted executives from both companies in New York to open negotiations. They departed with plans to talk again. Several of Great Western's board members, as well as some bankers, arrived in San Francisco in November on an overnight train, prepared to make a deal.

PG&E was older and more established than Great Western. But Great Western was in a position of strength. Already, its generation system was significantly more powerful than that of PG&E. It operated the powerhouse at Big Bend, as well as several steam plants in Oakland and San Francisco by virtue of its acquisition of City Light & Power. And the great dam was underway. Its share price reflected the anticipated value. A Great Western executive told reporters that his company had the upper hand in negotiations.

"The Great Western Power Company is not particularly anxious to merge with any other company," he said. "It is not making any advances and is not likely to accept any offers unless they are particularly attractive."

Talks between the companies broke down when PG&E put forth an offer. The company proposed acquiring Great Western for $15 million—half of what its directors were willing to accept. They packed up before Christmas and boarded the train back east.

PG&E's president at the time put an end to frenzied speculation about whether a merger would happen anytime soon. "Not only has the Pacific Gas and Electric Company not bought the Great Western," he told the *San Francisco Chronicle*, "but there are no negotiations and nothing but a fight between us."

———

Around this time, the power industry was awakening to the fact that electricity and gas providers could be considered "natural monopolies." It made little sense for competing companies to build duplicative power plants, wires, and pipelines, the thinking went, for that would diminish returns on capital-intensive investments. Samuel Insull, Edison's foremost protégé, was an early proponent of the idea. Insull, who happened to sit on PG&E's board of directors, was president of Commonwealth Edison, a massive power company in Chicago. He had effectively monopolized the market by acquiring every competitor in the vicinity. He began preaching the virtues of the electric monopoly as early as 1898, in a speech to the National Electric Light Association in Chicago.

"Acute competition necessarily frightens the investor, and compels corporations to pay a very high price for capital," he said. "The best service at the lowest possible price can only be obtained, certainly in connection with the industry with which we are identified, by exclusive control of a given territory being placed in the hands of one undertaking."

The notion that competition was counterproductive and risky gained traction in the following decade as power companies across the nation grew quickly by acquisition. And regulators were beginning to recognize the need to oversee them. The budding monopolies looked and acted a lot like the railroad companies that had grown to dominate their respective regions in the decades prior, setting arbitrary prices that customers had no choice but to pay. Regulatory commissions were created to ensure that the companies charged fair and equitable rates.

As electric and gas monopolies grew, states including New York and Wisconsin were among the first to appoint regulators to hold the companies accountable on price and service. Insull urged industry leaders to submit to greater oversight in exchange for protected profits and territories. "Our business is a natural monopoly and we must accept, with that advantage, the obligation which naturally follows," he said at a gathering in Chicago in 1910. "Namely, regulation."

In California, that regulatory authority initially fell to what was known as the Railroad Commission, which had been established to rein in the Southern Pacific Railroad, the state's most powerful monopoly at the time. In 1911, the state legislature expanded the commission's jurisdiction to include the fast-growing power companies cropping up throughout the state. The result was the Public Utilities Act of 1912, a book-length statute that gave the commission authority to regulate their rates, services, and operations. The commission would allow a power company to exist as a monopoly, but only if it provided affordable and reliable service to all customers throughout its territory. Otherwise, the commission might permit competition.

The statute reflected the views of John Eshleman, a lawyer and lawmaker with a scholarly bent. Eshleman envisioned the regulator as an arbiter working in the "best interest of the financier, who must advance capital, the corporation, who should utilize it, and the people at large, who must ultimately repay it." The Public Utilities Act, he wrote, was a "menace to those who desire to profit at the public expense."

Eshleman, who went by Jack among friends, grew up destitute in rural Illinois and fled to Southern California to find pick-and-shovel work in a railroad camp. He was said to have kept Aristotle in his back pocket. At night, he studied for college by light of a lantern. He passed the entrance test at Berkeley, where he steeped himself in philosophy, law, and economics. He left with two degrees and was then admitted to the bar. Tall and thin with small rounded glasses, he served briefly as a state lawmaker and district attorney before his appointment in 1911 to head the Railroad Commission.

Eshleman recognized the thorny challenge the agency would face as gas and electric companies consolidated. As monopolies, such companies could sell shares and issue debt more easily, allowing for larger investments in expansion and service improvements. But monopolies also have what's known as market power, the ability to set prices for goods and services at levels that wouldn't hold in the face of competition. As competition diminished, Eshleman foresaw the regulator stepping in to set not only fair and reasonable prices for service, but also rates of return on investments in new gas and electric infrastructure. The construct would serve to protect customers from price swings and all but guarantee profits for shareholders, allowing the utility to raise more capital.

"In short, we say to these agencies . . . government will only allow you what you ought to take," Eshleman once said. "And when we have said this, we have placed government in the position of determining the 'ought.'"

Just weeks after the commission's reorganization, Great Western applied for permission to sell power in Sonoma, Napa, Marin, and Solano Counties. Great Western had by then expanded its transmission network and planned to build a distribution system in Santa Rosa and other growing towns surrounding San Francisco in areas that PG&E already occupied. PG&E sought to block its competitor, arguing that its service was adequate.

Great Western countered by telling the commission that there were a number of underserved customers in the area, mainly in Santa Rosa, where it had discovered several hundred buildings without access to power. It said it could extend service to places where PG&E didn't operate, and offer lower prices across the region.

The commission agreed to hold a series of hearings in Santa Rosa and other cities to determine whether it should permit the competition. Great Western took out large newspaper ads encouraging residents to show up. "It is most important to the welfare of your community that you attend one of these meetings and give the commission your views," they read.

PG&E, for the first time, faced the real prospect of losing a substantial number of customers. It countered with its own ads, plastered across the pages of every local newspaper in Northern California. The company offered to fix unsatisfactory lighting systems, supply new lamps at cost, and perhaps even cut prices. "Tell us your needs, and we will do our best to satisfy them," one of the ads read. "'Pacific Service' is 'Perfect Service.'"

Dozens of people living in Santa Rosa and other nearby towns felt otherwise. They appeared at the hearings in support of Great Western. Santa Rosa's newly elected mayor took the stand to complain that PG&E had charged him thirty dollars to start service despite the fact that a power line had already been built to serve his neighbors. A resident on the outskirts of town explained his trouble getting PG&E to power his home, even after he offered to wire it himself. A chicken rancher from Petaluma showed up to say that he would happily buy Great Western's power, and that other chicken ranchers would, too.

The commission then heard from the companies themselves. Guy Earl appeared in his capacity as an attorney to represent the company he had helped build from the ground up. Charles Cutten, an attorney who had served for years in PG&E's legal department, argued on behalf of his company.

The two men took their seats in front of the commission one June morning. PG&E's president had arrived for cross-examination. The tension was palpable. Cutten became exasperated with Earl's questions and admonished him to pay closer attention to the executive's answers. Earl jumped to his feet. "I'll have you understand that neither you nor your corporation can crack any whip over me!" he yelled. Cutten sprang from his chair. Both men whipped off their glasses. Eshleman cried out from the dais as Earl drew a fist.

"Sit down, and apologize!" Eshleman exclaimed. "If this goes any further, we will take it as an excellent opportunity to test the power vested in this commission to punish for contempt of court."

The men mostly kept their tempers in check during the final hearing a few days later. Earl closed his arguments by stating that Great Western could sell power, at a profit, for just two cents a kilowatt-hour—a fraction of what some PG&E customers were paying.

Cutten responded snidely. "If the Great Western wants to come into this territory and reduce rates, we can do the same thing," he said. "We can lose money for as long as the Great Western can—and then some. And that's what we will both do if this competition is permitted."

He reminded the commission that power companies across the country were consolidating for a reason. "The big cities of the East are finding out that regulation, and not competition, is the best and only way to deal with public-service corporations," he said. "That is what we ask here."

Though the commission was beginning to recognize the value of consolidation, it decided that PG&E wasn't providing the sort of affordable and reliable service needed to warrant its protection as a monopoly. It granted Great Western the right to operate in most of the territories it sought to enter. PG&E would have to compete.

By the end of 1912, Great Western's charm campaign was in full swing. It had nearly completed renovating a new office building on Post Street in downtown San Francisco, in the heart of the shopping district. It transformed the first two floors of the six-story building into an "electric shop" with displays and demonstrations of the newest electrical appliances on the market—vacuums, irons, washing machines. All were for sale. Part of the second floor was specially designed for women, with a powder room, telephones, and desks stacked with writing materials. Visitors who climbed to the roof found a garden and a sunroom with a small café. Wicker chairs sat beneath a pergola adorned with English ivy and geraniums.

The company had also nearly finished the dam at Big Meadows, where it had acquired thirty thousand acres of land. Come spring of 1914, the reservoir began filling rapidly with rain and snowmelt, forming a lake framed by mountains and lush forests. Though it was Howells's best-known accomplishment, he named it not for himself, but for Guy Earl. Howells called it Lake Almanor, an amalgamation of Earl's daughters' names—Alice, Martha, and Elinore.

Great Western then set to work on completing the system. It built another powerhouse, called Caribou, on a curve in the river fifty miles upstream from the one at Big Bend. It extended its transmission line through the remote reaches of the Feather River Canyon, with steel towers perched along rocky outcrops. By 1924, Great Western delivered power to about 1.4 million people across five thousand square miles. That year, in a solicitation to sell more shares, it alluded to plans to expand even further. Profits, it said, were all but guaranteed, for the provision of power was an essential need in a state where growth seemed boundless.

Howells's life's work was complete. His early investment in Great Western had made him a wealthy man. He drove a Cartercar, which looked like a motorized horse carriage, and had a home on the ritzy Belvedere Island overlooking San Francisco. His daughter got married in the garden there on a summer evening in 1923. He became a grandfather the following fall when his son's wife gave birth to a girl. Howells died just a few years later, in 1927, at age sixty-eight. He had a severe stroke during a Commonwealth Club luncheon in San Francisco.

It was the end of an era. Howells's fellow electric pioneers were aging, too. Some had already passed—Schuyler died in 1912. Two years later, Martin and de Sabla relinquished their seats on the board of PG&E to pursue new ventures in semiretirement. Martin died in 1928, while de Sabla moved to New York and dabbled in other industries.

With that, Great Western and PG&E each passed into the hands of a younger generation. James B. Black, a young Californian who had become a Great Western service inspector right out of college, would be the one to unite them. Within a decade, Black had climbed the ranks to become an executive just below Guy Earl. Under their leadership, Great Western grew even larger by acquiring nearly every remaining power company in the region. Then, in 1925, Great Western itself was acquired by the North American Company, which owned several utilities across the country. Black moved to New York and became the company's vice president.

Unlike Earl, Black hadn't lived through the power industry's competitive heyday. He hadn't tried to start a fistfight during a Railroad Commission hearing. By the time Black joined Great Western, the utility business was heavily regulated and becoming more monopolistic by the day. Great Western, meanwhile, was still in fierce competition with PG&E, forced to keep prices low while also collecting enough revenue to invest in expansion. Perhaps more than Earl, Black recognized the value of merging Great Western with PG&E to form one of the largest utility companies in the nation with the exclusive right to serve customers in Northern California.

Black traveled from New York to San Francisco to start negotiations with PG&E in 1929. The following year, he struck a deal for a stock swap: North American would give PG&E its shares in Great Western, and in exchange, North American would acquire shares in PG&E. It was a win for both companies. PG&E would at last acquire its only remaining competitor, while North American, would reap returns from its investment in a larger and more valuable utility. Black later stated that the acquisition "removed the hazards of competition" and "rounded out the Pacific Gas group into a large system that would be very economically operated."

The Railroad Commission approved the merger on May 14, 1930. It had articulated its support for such consolidation years earlier, in a series of decisions that protected monopolies in various industries throughout the state. "A utility which secures all the business in a certain territory is much better able to give good service at reasonable rates and to make the necessary extensions than though its revenues are diminished through competition," the commissioners wrote.

"It is a well understood maxim that in cases of competition in public utility service, the public, in the long run, generally pays the bills, including the cost of all duplication and other economic waste."

In approving the merger, the Railroad Commission became solely responsible for determining PG&E's rates of service and returns—just the sort of arbiter John Eshleman had envisioned when the Public Utilities Act was enacted in 1912. When Eshleman died a few years later, Earl delivered his eulogy. Eshleman's work heading the Railroad Commission, though brief, helped establish the framework that would allow utilities to become some of the nation's largest, and most enduring, monopolies. "He became widely known throughout the country," Earl said. "Through his constructive ability and supreme courage, the great work of that commission was begun and given its impulse."

Both Black and Earl joined PG&E's board, with Black becoming its president in 1935 at age forty-five. He sat straight-faced for a photo, dark hair combed and parted on the left, brown eyes shadowed by thick brows. Earl died that year, at seventy-four, closing the chapter on Great Western's origin story. In the thirty years since Julius Howells had pitched Earl on flooding Big Meadows, Great Western had built or acquired more than a dozen hydroelectric powerhouses, eight steam plants, and a large network of power lines, much of which traversed the dense forests of the Sierras. PG&E executives liked the Caribou powerhouse so much that they began hosting guests at the clubhouse there, accessible by a company-operated railway that ran nine miles from a tiny mountain town.

As anticipated, the merger enabled PG&E to invest heavily in growth. It first turned its attention to building a pipeline system to deliver natural gas to Northern California, where the resource was scarce. Then, after World War II, the company plowed more than a billion dollars into the expansion of the electric system. The state had seen a massive influx of people to support wartime industries, and demand for electric service skyrocketed as they bought homes and started families. Within six years, PG&E had roughly doubled its capacity to produce power.

In July 1952, PG&E was awarded the Charles A. Coffin Award, named for a cofounder of General Electric. It was the utility industry's highest honor at the time. The judges acknowledged PG&E for completing what they called the most ambitious expansion effort ever undertaken in the industry. The company had "dramatically demonstrated its ability and willingness to meet all needs for electric power," the judges wrote.

A photographer at the award ceremony snapped a photo of Black, whose dark hair had gone white. This time, he beamed as he accepted a gold medal embossed with Coffin's bust, the ultimate recognition of his seventeen-year tenure as PG&E's president. He had successfully secured PG&E's foothold in almost all of Northern and central California. Its acquisition of Great Western had been key to that success, giving PG&E the size and financial wherewithal to invest in a system of power plants, wires, and pipelines spanning seventy thousand square miles.

But the acquisition also seeded a problem that would take decades to germinate. PG&E was now responsible for a vast network of transmission lines that it had no role in building. And many of the records and correspondences the company inherited would eventually get lost or purged with time.

Decades earlier, at the turn of the century, Great Western's founders had weighed two options for the transmission line that would eventually travel from the Caribou powerhouse to Butte County. They considered building it to traverse the snowy Sierra peaks in a direct line toward Chico, which would require cutting thousands of trees to create the necessary right-of-way. An alternative path would follow the remote expanses of the Feather River Canyon, where the trees weren't as dense. Schuyler, the engineer, had opted for the latter route, though he acknowledged the challenges of accessing it. "The railroads of the future, which are certain to be built into this region to tap the great timber resources, will inevitably follow the canyon," he wrote. "This will facilitate the patrol and repair of the lines."

Great Western purchased some of the transmission hardware from Ohio Brass, including metal hooks to hang the wires and insulators along the Feather River Canyon. Like the power line itself, the hooks were somewhat experimental—they were made of malleable cast iron, which, it was soon discovered, lost integrity with age and exposure to the elements. In 1924, three years after the Caribou-Palermo was complete, Ohio Brass switched to a new manufacturing method that produced iron "freed from all tendency toward embrittlement" when processed a certain way. It also began offering forged steel components, which were stronger still.

Nearly a century later, one of Great Western's malleable cast-iron hooks gave way, dropping a transmission wire that started a wildfire off a gravel road in the Feather River Canyon, where a train runs only occasionally. Investigators asked PG&E to hand over its records on the origin of the hook, as well as the transmission tower and the wire that fell. The company had almost nothing to give.

Three

DEATH MARCH

During a sleepless few weeks in the summer of 1996, a California state senator named Steve Peace led what became known as his "death march," an exhausting, caffeine-fueled legislative marathon. It began in July in a hearing room deep within the pillared capitol building in Sacramento. Peace, a trim man with heavy eyebrows, round glasses, and thick brown hair receding past his forehead, sat in the center of the dais beneath a wall-to-wall panorama depicting California's rich history. His task: to roll back the decades-old rules allowing utilities to function as monopolies, and reintroduce competition in the provision of electricity throughout the state. The process was called deregulation.

It was an enormously complicated undertaking that would have bored or frustrated most anyone. But Peace, a Democrat from San Diego, thrived where others suffered. As a young member of the California assembly, and then the senate, Peace developed a reputation for diving deep into subjects that few people cared to understand. Electricity was one of them. Peace had buried himself in the mounds of documents that spelled the rules for California's three large utilities—PG&E, Southern California Edison, and San Diego Gas & Electric—with the intention of breaking apart their legacy businesses.

There were other things that set Peace apart in the halls of the capitol. For one, he had cowritten and starred in the cult classic *Attack of the Killer Tomatoes!*, a satirical, low-budget film that makes fun of bureaucracy and government ineptitude. The movie came out in 1978, when he was twenty-five. Four years later, he was elected to the assembly, making him the youngest member of the California legislature. Despite his youth, he spoke his mind. He was, by his own admission, something of an egomaniac. But he was smart, and he worked hard. He didn't suffer fools.

When Peace became head of the state's Senate Committee on Energy in 1995, the United States was emerging from a deep recession that had been particularly acute in California. Between 1990 and 1993, the state lost more than half a million jobs as aerospace and defense companies downsized in the wake of the Cold War and a booming real estate market reversed course. The recession was a result of a number of complex factors stemming from the state's rapid growth in the prior decade. But it served to highlight one particular problem: Californians were paying exceptionally high prices for electricity. In 1991, the average rates charged by the state's utilities were as much as 50 percent above the national average. That gave way to calls for deregulation.

Peace, with his penchant for complexity, would be the one to carry the mantle. His death march began with a summons. Anyone and everyone who wanted to buy or sell electricity in a new competitive market would appear in Sacramento to debate how and when to jettison old regulations. Peace channeled supreme confidence as he outlined his intent for the discussions, becoming almost giddy as he raised hypothetical questions about the role of the utility and who should pay for the "mistakes" of the past. He had clearly thought longer and harder about deregulation than almost anyone else in the room.

Peace set the stakes during an afternoon session in mid-July. Wearing a dark blue suit jacket and a patterned purple tie, he admonished a group of industry representatives for thinking inside the box. The debate, he said, wasn't simply about restructuring the electricity business. It was about rehabilitating California's economy. As Peace saw it, the state's future was in their hands.

"I see potential for dramatic benefit to the average Californian in the form of a higher paycheck, in the form of a more secure economic environment in general, and in the form of an entire generation of Californians that have a higher level of expectation that maybe their kids can actually stay and live here when they become adults, because there might be a job for them," he said.

The impetus for electricity deregulation began with the dashed promise of nuclear energy as a cheap, plentiful source of power, as well as a series of crises that caused oil prices to skyrocket. Together, those dynamics had major consequences for utilities and their customers and caused regulators across the country to question whether monopolies born of the early twentieth century could hold up in an industry that had changed significantly in the intervening decades.

The shift began shortly after World War II, when the US government established the Atomic Energy Commission and tasked the new civilian agency with finding peacetime uses for nuclear fission. Fission, a self-perpetuating chain reaction, occurs when an infinitesimal particle splits the heart of a radioactive atom, generating energy that propels other particles to split other atoms. Left unchecked, the reaction accelerates exponentially and can cause explosions like the ones that leveled Hiroshima and Nagasaki. It can be used to create, rather than destroy, but only if it's tightly controlled.

The reaction generates heat—more than enough to boil water to power steam turbines, which at the time were fueled by coal, oil, or natural gas. Scientists theorized that fission, too, could be used for electricity generation. An experimental reactor in Idaho demonstrated the concept in 1951. Three years later, the chair of the Atomic Energy Commission addressed members of the National Association of Science Writers at a convention in New York City. He implored them to help the public understand the atom's benevolent potential. "It is not too much to expect that our children will enjoy in their homes electrical energy too cheap to meter," he said.

First, the commission would have to convince utilities across the country to help develop nuclear reactors. The federal government would help fund the cost of research, but not construction. That threatened to put utility customers on the hook for big investments in power plants far riskier than those that ran on fossil fuels. The technology was young and inherently dangerous. Every reactor would require serious safeguards to protect employees and the public from radiation exposure. And there wasn't yet a commercial market for nuclear fuel.

A number of utilities saw benefits alongside the risks. Nationally, electricity demand had been growing by nearly 10 percent a year on average, threatening to exceed supply in some places. Nuclear plants could go a long way to close the potential gap. They could also help to hedge against the uncertainties that came with purchasing oil from foreign producers. The United States relied on the Middle East for as much as a third of its supplies, some of which was needed to fuel power plants. Price spikes or supply disruptions could drive up electricity costs.

Utilities in California, which was home to several nuclear research sites, were among the first to bet on the technology's potential. In the mid-1960s, Southern California Edison and San Diego Gas & Electric collaborated to develop the San Onofre Nuclear Generating Station just outside of Los Angeles, while PG&E planned the Diablo Canyon Power Plant near San Luis Obispo. Each plant was expected to cost roughly half a billion dollars to build.

The first reactor at San Onofre, contained beneath a giant steel sphere on the edge of a popular beach near San Diego, began producing power in 1968. Politicians and scientists heralded it as a symbol of the peace atom during its commissioning. San Diego Gas & Electric's president at the time reminded an audience of about four hundred that it had taken less than a century for the utility industry to harness one of the most powerful forces on Earth.

"The electric light was invented as recently as 1879—within the lifetime of a few people alive today," he said. "Nuclear power has had an even more amazing development."

But a run-up in the cost of producing electricity, whether with fission or fossil fuels, was about to cast the industry into turmoil. In 1973, the largest oil-producing countries in the Middle East sent prices soaring by slashing production and halting exports to the United States in retaliation for its intervention in the Arab-Israeli war. Fuel shortages came fast and swift. Volkswagen Beetles and Ford Pintos queued up at gas stations forced to ration their supplies.

It wasn't just gasoline. Oil-fired power plants suddenly cost far more to operate. Gas-fired ones did, too, amid a natural gas shortage exacerbated by the oil crisis. Utility customers would soon foot the bill. President Richard Nixon signed a law to permanently extend daylight saving time in a bid to reduce evening electricity use, only to roll it back when children were left standing in the dark as they waited for their morning school buses.

Congress, desperate to reduce the nation's reliance on the Middle East, passed the Public Utility Regulatory Policies Act, a 1978 law meant to stimulate domestic energy production. The law seeded a nascent wholesale electricity market by encouraging companies other than utilities to build wind and solar farms, or gas-fired plants that also produce usable heat in a process called cogeneration. Those technologies, expensive at the time, had little hope of competing on price with conventional power plants. So the law required utilities to buy electricity from the newcomers for roughly what it would have cost to produce it themselves using oil, or, in some cases, other fuels. A sharp drop in oil production during the Iranian Revolution again sent prices soaring in 1979. The utilities signed long-term contracts with the new wholesalers on the assumption that cheap oil was gone for good.

That same year, mechanisms used to cool a nuclear reactor at the Three Mile Island plant in Pennsylvania malfunctioned, causing the core to overheat to the point of meltdown. Plant operators managed to contain the worst of the fallout, but the accident nevertheless had a chilling effect on the nuclear power sector, which was already grappling with cost overruns caused by inflation, regulation, and public skepticism.

San Diego Gas & Electric and Southern California Edison, which had decided to expand San Onofre with two more reactors, faced major delays, in part because regulators had imposed new safety requirements. The plant ultimately cost about $4.5 billion when it was completed in 1984. PG&E faced similar challenges in completing Diablo Canyon. When the company at last finished it in 1985, PG&E had spent more than $5 billion.

The overruns posed a major challenge for the California Public Utilities Commission, or CPUC, which, in a sign of the times, had shed its name as the Railroad Commission in 1946. Historically, it had allowed utilities to recoup the costs of building new power plants and earn a return on their investments by raising rates. But these power plants were among the most expensive the state had ever seen, and the utilities had seriously strained their balance sheets in building them. The question was who would have to bear the costs, shareholders or customers? Forcing investors to take a hit could sap the utilities' financial strength, jeopardizing their ability to make future investments. Asking customers to pay more also had economic consequences.

In the case of San Onofre, the CPUC allowed Southern California Edison and San Diego Gas & Electric to bill for most of the construction costs, though their investors ate a portion. PG&E faced a different situation: The commission tied the company's reimbursement to Diablo Canyon's performance. Customers would pay for only the electricity it generated. That power, however, was relatively expensive, and the plant produced it in abundance.

In the mid-1980s, oil prices declined and stabilized. Supplies swung from deficit to surplus as countries around the world cut back on consumption. In the United States, utilities had built more power plants to use coal or natural gas instead of oil, and focused on energy efficiency. The problem was that the pricey contracts they had signed for wind, solar, and cogeneration—based on the assumption that oil prices would remain high—suddenly appeared unduly expensive.

Electricity from those sources now cost more than the power the utilities could produce themselves. And customers were once again stuck with the bill.

At the time, California's economy was booming. It was easy to overlook the inexorable rise in electricity prices as inflation rates stabilized and jobs multiplied. But in the early 1990s, during the sharp and painful contraction, utility customers—especially large power users—scrutinized their bills. Some businesses threatened to leave the state. That didn't sit well with the CPUC. The agency had for years been mulling whether opening California's electricity industry to competition would result in lower prices. It began to consider the prospect of deregulation in earnest.

Put simply, deregulation is the process of relieving the regulator of its role in setting prices for a given service and instead handing that function to a competitive market. Already, several monopolistic industries had been deregulated in the 1970s and 1980s. Telephone customers suddenly had a choice among service providers after the government-ordered breakup of AT&T's national network. Railroads, airlines, and natural gas suppliers also faced new competition. Momentum was building to apply the same principles to the provision of electricity. The Energy Policy Act, a sweeping federal law enacted in 1992, made it easier for companies other than utilities to build and operate power plants, undermining the monopoly model for generating electricity. Regulators in states across the country began debating how to allow for competition among the newcomers.

Deregulating the electricity business meant that utilities would, for the most part, no longer produce power and sell it to customers.

Those functions would instead be assigned to wholesale suppliers, which would run power plants and sell the output in bulk, and retail suppliers, which would buy that bulk power and parcel it out to homes and businesses. Utilities would essentially become intermediaries tasked with maintaining the infrastructure needed to move electricity around.

California was among the first states to initiate the deregulation process. In a thick 1993 report, CPUC staffers wrote that the so-called regulatory compact, the longstanding agreement allowing utilities to function as protected monopolies, no longer worked as well as it once did, when the electricity business was simpler. Regional utilities had become subject to global market forces. Their means of producing power had evolved and multiplied. The CPUC, over several decades, had taken increasingly drastic measures to help the utilities manage a new set of risks to their shareholders, often at the expense of their customers in the form of higher prices. It was time, the staffers wrote, to rethink the nature of the compact. In other words, the notion of the natural monopoly was suddenly up for debate. So, too, was the role of the regulator.

Deregulation was a bit of a misnomer—in the electricity business, regulation can never be fully eliminated. In allowing for competition, regulators would have to keep an eye on electricity markets and determine how utilities should be compensated for power line upkeep. But they would no longer have nearly the same say in determining what electricity ought to cost, a pivot from the philosophy John Eshleman had put forth in the Public Utilities Act of 1912, when California utilities were first becoming vertically integrated monopolies. Electricity would become a commodity, its price dictated by levels of supply and demand.

The CPUC president at the time was willing to relinquish much of the agency's regulatory control. The forces of competition, he wrote, "represent a superior disciplinary force than can be supplied over time by any system of governmental estimate or planning of which I am aware." The commission proceeded to roll back some of the regulations it had put in place decades earlier. It ultimately outlined a set of steps to break the utilities' hold on power generation and sales to customers. It then handed the task to the California legislature.

—

As the CPUC debated deregulation, California's three big utilities resigned themselves to change. They saw there was no use in fighting for the fractured status quo. While they didn't uniformly like the concept of competition and had different ideas for how it should work, they recognized potential opportunity. Deregulation would allow the utilities themselves to establish unregulated businesses to play in new markets.

PG&E's chief operating officer at the time was Bob Glynn, a mechanical and nuclear engineer who would soon rise to become the company's CEO. He had gotten his start in the utility industry well before the passage of the 1978 law that seeded the wholesale power market. PG&E hired him amid the turmoil of the 1980s, and he rose through the ranks of the company's electric division. By the time he became an executive in 1994, the industry was on the cusp of transformation. He appeared that year before a subcommittee of the US House of Representatives to discuss the future of the regulated utility. PG&E, Glynn said, understood that deregulation was inevitable and supported the idea—so long as the construct benefited both its customers and its shareholders.

"People speak as if restructuring of the electric industry is a future event to be contemplated, analyzed, debated," Glynn said. "That is pretty far from the truth."

Already, PG&E was transforming itself to survive in a new competitive environment. It announced plans to cut three thousand employees—roughly 11 percent of its workforce—by the end of 1995. It began closing local offices serving customers throughout its service territory, centralizing operations in preparation for a major reorganization of the business. It created PG&E Corporation, a holding company that would soon become the umbrella for Pacific Gas and Electric Co., the regulated utility, as well as several unregulated businesses that would generate electricity and transport gas not just in the United States, but in other markets overseas.

California's other utilities were also cutting jobs and reining in spending in preparation for deregulation. Their acquiescence came at a price. The nuclear plants they had spent billions of dollars building had no hope of competing with cheap fossil fuels. And the expensive contracts stemming from the 1978 law were still on the books. So the utilities lobbied hard to continue recouping those costs, arguing that they needed the money to survive.

In truth, Steve Peace wasn't totally comfortable with the idea of deregulation when it came time to implement it. He recognized the risk in trading price stability for price volatility in the provision of a critical service. But he had been assigned the task because he had the mind for it. He was committed to achieving the objective in a way that protected residential customers from major price swings and created a transparent market for trading power. He held the first hearing on July 11, 1996, and then another in early August. So began the death march.

For weeks, Peace worked late into the night with attorneys and lobbyists, regulatory staffers and consumer advocates. The hearing room often remained packed well past midnight. More than once, in the early hours of the morning, Peace yawned and stretched, signaling that he was ready to wrap it up. His soldiers began packing their briefcases. Then Peace would walk to the coffee machine and grab a cup, as well as a bag of M&M's, and return to the dais. The groans were audible.

Peace was a smart-ass, and a funny one at that. His sarcasm often got people laughing. But negotiations were often tense, if not acrimonious. When parties disagreed, Peace expelled them from the hearing room and wouldn't allow them to return until they had reached a compromise. On the night of August 10, the parties were tired. As usual, it had been a long night, with negotiations still going at 10:45 p.m. Peace, who had forgone a suit for a green short-sleeved polo shirt, reached his breaking point during a meandering disagreement between two men, one representing PG&E and the other its agricultural customers. Peace threatened to resolve the issue with no input from either party.

"I'm not interested in negotiating between the two of you anymore," Peace told the representatives. "Don't talk to us again until you can quantify what we're talking about."

"Excuse me," one of the representatives ventured.

"I'm done!" Peace snapped, holding up his hand.

The men hustled out of the hearing room.

A senator seated to Peace's left looked to the center of the dais. "Have a good night's sleep?" he said with a smirk.

"Yeah, I'm in great shape," Peace retorted.

It took another two weeks to hammer out a piece of legislation. The final draft ran nearly sixty pages. On August 24, the parties spent from 10:00 a.m. to 8:30 p.m. reviewing the bill line by line. The next day, Peace told them that copies were, at last, being printed.

"It's been the most scrutinized document in the history of the legislature," Peace said. "It's also probably going to be the first legislative package ever passed that's longer than the budget, so it ought to be."

The bill was designed to create a world in which electricity customers were no longer beholden to their utilities and could instead choose among a number of retail suppliers—exactly what California's big businesses had been pushing for at the height of the recession. The transition would occur in phases. At first, competition would be mostly limited to a new market called the California Power Exchange, where wholesale suppliers would sell electricity to the utilities for retail distribution. Utilities would remain the state's primary retailers until about 2002, when customers were expected to begin shopping among new suppliers. At that point, the utilities would essentially become intermediaries tasked only with the transport of power.

To facilitate this, PG&E, Southern California Edison, and San Diego Gas & Electric would keep their hydroelectric and nuclear plants but sell the rest to suppliers that would compete within the Power Exchange. The utilities would still own their transmission lines, but the suppliers would have access to them in order to route power deliveries. A new nonprofit authority called the California Independent System Operator would oversee traffic on the grid and monitor levels of supply and demand.

The deregulation bill passed in September 1996 without a single dissenting vote. With that, the utilities were transformed. Transmission lines, their private highway networks, had essentially been opened to the public. And the utilities would no longer produce most of the electricity they sold to customers. Large power companies including AES, NRG Energy, Mirant, and Duke Energy purchased their plants, and a slew of others, including a brazen Texas firm called Enron Corp., set up shop to trade in the new market.

The utilities had notched several wins in crafting the legislation. Most significantly, they had helped design a rate structure that would allow them to bill customers for their financial albatrosses, formally called "stranded costs." The legislation implemented a freeze on retail rates, shielding customers from price swings as the new markets were established. Competition was expected to push wholesale power prices well below retail ones. The law permitted the utilities to pocket the difference until they were made whole. In the interim, rates would remain frozen, and utilities would not be permitted to charge more to their customers. That prohibition was the piece of the puzzle that doomed the whole experiment.

Four

MARKET FAILURE

When George Sladoje arrived in California from Chicago in 1997, he didn't know he'd be a short timer. He had been hired to help build the Power Exchange, the cornerstone of the state's new deregulated market. The Power Exchange would serve as the platform allowing wholesale suppliers to sell electricity as a commodity. Sladoje understood markets, which facilitated the buying and selling of all types of commodities around the world. He had most recently been chief financial officer at the Chicago Stock Exchange and had worked for more than a decade at the Chicago Board of Trade. But he knew nothing about electricity, a far cry from the agricultural and industrial products that flashed on the screens in downtown Chicago. His job was to help commoditize it for trading across the West. It was a hugely important role, and it excited him. He saw it as a chance to make history and began studying up.

Building an electricity market is a daunting challenge. Unlike physical commodities such as oil and gas, electricity is subject to a special set of constraints. Most critically, supply and demand levels must be kept in constant equilibrium. If demand exceeds supply by even a small amount, wide swaths of the grid become vulnerable to failure because of the way the system is calibrated. Maintaining the balance is a challenge because electricity is generally consumed the moment it's produced. Storing it has, until recently, been difficult and expensive, so there is not a meaningful surplus held in reserve for use during supply crunches. And transmission lines, the high-voltage highways traversing long distances, can handle only a certain amount of electricity. They become congested if too much is loaded onto any one conduit, potentially tipping the supply-demand balance in the same way a blocked shipping channel disrupts supply chains. In order for the grid to function smoothly, power plants must produce exactly enough electricity to meet demand, and that electricity must travel unimpeded along transmission lines connecting those plants with cities and towns.

As part of the deregulation push, California's utilities ceded control of the two key variables in that process—power plants and transmission lines—allowing for the creation of the Power Exchange. Despite its unique constraints, the exchange would function much like any other commodity market. Just as oil is sold in barrels and natural gas is sold in British thermal units, electricity would be sold in big blocks called megawatt hours, one of which equaled the amount of power needed to supply several hundred homes for an hour. Companies called generators would operate power plants and sell their output at various hubs on the transmission system, where traders could transact for that power with the intent to sell it at a profit either among themselves or to the utilities through the Power Exchange. More so than the generators, the traders would become responsible for routing the movement of electricity between different hubs, where they could sell it at prices determined by local supply, demand, transmission costs, and a slew of other variables. At its simplest, trading is a game of arbitrage: buy low, sell high.

The Power Exchange set up shop in an office suite about ten miles outside the heart of Los Angeles. Sladoje and his staff lined a small room with boxy desktop monitors and threaded power cords behind a row of desks. It had none of the frenzy of an open trading floor, but it would serve a similar clearinghouse function. Sladoje helped establish a "day-ahead" market, an auction where utilities would buy the power they expected to need during each hour of the following day. The exchange would assess bids from suppliers and determine hourly prices.

The process was straightforward. The Power Exchange would calculate total expected demand in any given hour and sort supply bids by price, from lowest to highest. It would first accept the lowest-priced bid and work up the list until it had secured enough supply to satisfy demand. The most expensive bid accepted through that process set what's known as the market clearing price, the amount paid to any supplier who offered to sell at or under that threshold. Any bid above that threshold would be rejected as unnecessary.

The Power Exchange would then relay the schedules to the grid operator, which would oversee power delivery and monitor supply and demand in real time. Balance was critical, given that even a brief mismatch could cause widespread blackouts. To account for last-minute fluctuations, the grid operator would run an auction where it could purchase emergency supplies at almost any price the market dictated. It was designed as a backstop, and hardly anyone thought the grid operator would have much reason to use it.

Setting up the Power Exchange took months. Sladoje could hardly stand all the meetings. Staff of the grid operator, the CPUC, the legislature, and the utilities bickered almost every week about how the market should function. It was supposed to open in January 1998. Come the New Year, it was nowhere near ready. It wasn't for lack of effort. The Power Exchange had a small but mighty staff that worked long hours to hone the market design. They ran simulations in the spring, only to find the software full of glitches. The system kept crashing. It took weeks to scare out the bugs. The pieces finally fell into place late one Sunday night in March. Sladoje made some calls. It was ready.

The Power Exchange began processing trades on the last day of the month. Generators and traders bid supplies into the market, and California's three big utilities each placed their orders. The exchange cleared the trades and shipped the schedule off to the grid operator, where staffers celebrated with a lightbulb-shaped cake. All told, $10 million worth of electricity changed hands.

A curious thing happened on that first day of trading: Prices came in much lower than expected. For a couple of hours, they fell to zero. Electricity was literally free. Sladoje's team had never encountered that scenario while running simulations. As it turned out, there was more than enough power to go around.

It was a sign of low prices to come. During the market's first year, the price of wholesale power averaged about $20 or $30 a megawatt hour—roughly half of what the utilities charged customers. The gap between wholesale and retail prices was so substantial that the utilities were recouping their bad investment costs more quickly than anticipated. Peace wrote a letter to update his fellow lawmakers on the new market, pointing to its success as proof that they had been right to introduce competition. The letter noted that prices a day earlier had been just $13.55 per megawatt hour.

"The success of the PX confirms the choices the Legislature made in adopting its customer-friendly model of electricity industry restructuring," the letter read.

For PG&E, the new market had opened the door to growth. The company's share price was on the rise, reaching levels roughly on par with where it had been earlier in the decade, before talk of deregulation spooked the market. It was the era of the "PG&E Shareholders" hat—cream-colored baseball caps that bore those words alongside an embroidered logo. The company sent them to investors as it worked to position itself as a stock to watch. (Executives had to laugh when an employee reported driving through Stockton and spotting a panhandler in one of the hats.)

Politicians and regulators in other states watched with interest. It seemed that California had blazed a path to successful deregulation. The grid operator's counterpart in New England, which oversaw that regional system, consulted with Sladoje about setting up a power exchange there. A similar call came from Alberta. But the California market wasn't nearly as strong as it appeared. Weak spots were fast emerging. Surges in demand caused prices to spike, resulting in a payday for every generator that put power up for auction. In the summer of 1999, the Power Exchange introduced contracts that allowed the utilities to buy power in the future at set prices, hedging their exposure to market swings. But the CPUC, recalling the missteps of the 1970s and 1980s, was apprehensive that those contracts might prove expensive in hindsight. It limited the utilities from straying too far outside the spot market.

The generators and traders were beginning to recognize how easy it was to tip the balance between supply and demand. Enron Corp. was chief among the traders. The Houston-based energy company had transformed, in a matter of years, from a sleepy pipeline operator into one of the most valuable companies on Wall Street. It had purchased Portland General Electric in Oregon in 1997 and set up a trading division there to focus on California's power market.

Heading the division was Tim Belden, a trader in his early thirties who had proven himself as a sharp energy analyst. He began his career as a researcher at the Lawrence Berkeley National Laboratory, a federal energy lab run by the University of California. He wound up studying electricity markets and coauthoring papers on headache-inducing topics such as electricity futures and price volatility. Like Peace, Belden had an uncanny knack for parsing the impenetrable.

Belden joined Enron in 1997, just before California's power market opened to competition. Enron traders, many of whom were Ivy League MBAs, had a reputation for being ruthless climbers. Their training materials were found to have included quotes from Louis Winthorpe III, the Wall Street investor character in the 1983 film *Trading Places*. ("In this building, it's either kill or be killed. You make no friends in the pits and you take no prisoners.") Belden didn't fit that mold. He was passionate about the environment, often biking to work instead of driving his old beater. He was collegial and occasionally goofy. But he was also shrewd and relentless in his own way, working long hours to crack California's new market structure. He recognized the loopholes almost immediately.

Some were simple. If suppliers withheld electricity from the Power Exchange, prices would go up. More complicated was the matter of transmission-line congestion. Because the utilities had ceded control of the grid, traders were free to route deliveries as they pleased, with little regard for capacity limits. The grid operator offered payments to those who could route power elsewhere to reduce system gridlock. Profits awaited those who deliberately created bottlenecks and then offered to relieve them. And finally, there was the tempting prospect of shorting the grid in real time. If a generator shut down a power plant that had already agreed to deliver power at that moment, the grid operator would pay huge amounts to replace the supply by holding an emergency auction, the last-ditch option that few had thought would be necessary.

All of this gave generators and traders market power, the sort of price-setting ability usually kept in check with regulation or antitrust laws. In theory, competition prevents market manipulation. But this wasn't a perfectly competitive market. Its peculiarities negated its safeguards.

One of Belden's earliest experiments occurred in May 1999, when he bid to sell 2,900 megawatts of electricity to the Power Exchange and scheduled it to move along a transmission line connecting Nevada and Southern California. It was a huge amount of power, enough to supply a midsize city, and there was no way to transport it along Belden's chosen line, which could carry only 15 megawatts. The trick created scarcity because the power couldn't reach California as planned, forcing the grid operator to buy replacement power at exorbitant prices and ultimately pay Enron to relieve the congestion that it had deliberately created.

The grid operator was immediately suspicious of the trade, though there wasn't much it could do about it. Its policing power was limited. A staffer phoned Belden shortly after he submitted the bid to ask why he appeared to have scheduled an impossible delivery.

"We did it because we wanted to do it," he replied. "I don't mean to be coy."

"It's a pretty interesting schedule," the staffer replied.

"It makes the eyes pop, doesn't it?" Belden asked.

"Um, yeah, I'll probably have to turn it in 'cause it's so odd," the staffer said, implying that she would alert the Power Exchange.

Sladoje, too, sensed something was amiss. That sort of overscheduling, he thought, violated the spirit, if not the letter, of the new marketplace rules. The Power Exchange launched an investigation to determine what ought to be done about it, just as Enron was gearing up to expand its role in the market. At the end of the year, Sladoje received a form letter signed by Ken Lay, Enron's chief executive, outlining the company's "commitment to conducting itself in accordance with the highest ethical standards." Enron expected the same of its business partners, the letter read, so that its employees could avoid "potentially embarrassing or unethical situations."

For a host of reasons, power became much pricier in the spring of 2000. Lack of supply was chief among them. California's population and economy had grown, but very few power plants had been constructed since the start of the decade amid uncertainty about deregulation. On top of that, a dry winter reduced hydroelectric generation in Northern California and the Pacific Northwest. And natural gas prices were on the rise, making power more expensive to produce.

In May, power sold on the exchange topped fifty dollars per megawatt hour, a record high. That number tripled the following month. In the southernmost corner of the state, the consequences were immediate. Months earlier, San Diego Gas & Electric had successfully recovered its stranded costs and lifted the freeze on retail electricity rates, exposing its customers to the volatile price of electricity sold through the exchange.

The summer was blazingly hot. Across California and the rest of the West, people blasted their air-conditioning units, straining a system already in short supply of power and driving prices higher still.

In San Diego, residents opened their bills to find they owed double or even triple what they had paid the previous summer. State lawmakers moved quickly to refreeze their rates.

For Pacific Gas and Electric and Southern California Edison, the rate freeze was devastating. They were prohibited from charging more than about sixty-five dollars a megawatt hour, a comfortable threshold in the era of low wholesale prices. But that summer, the honeymoon period came to an abrupt and painful halt. Bulk power cost anywhere from $150 to $1,000 a megawatt hour, and the utilities had no choice but to sell it for far less. Their debts mounted, sometimes by as much as $50 million per day.

The first outages occurred in June of that year as a heat wave swept the state. The grid operator faced the prospect of demand surging beyond available supply, straining the grid to the point of collapse. It ordered PG&E to cut power in phases to tens of thousands of customers throughout the Bay Area to help relieve the stress. PG&E rotated the outages, called rolling blackouts, throughout the region on an hourly basis in an effort to make sure no customer was in the dark for too long.

Bob Glynn, who had by then become PG&E's CEO and chairman, was angry and frustrated. The deregulation effort he had spoken in favor of was so broken that the company was forced to turn out the lights on its customers to keep demand in line with supply. A mismatch could cause the grid's operating frequency to drop and potentially cause serious damage to power plants and other equipment tied into the network. In that case, the blackouts would last not for hours, but for weeks.

There could hardly have been a better time to exercise market power. Almost every company hawking electrons played some role in price gouging. A number of generators were found to have feigned power scarcity by withholding supplies during the Power Exchange's daily auctions. Enron, however, had devised the most colorful and sophisticated means of gaming the system. Its traders employed a suite of complex schemes with names like "Death Star" and "Get Shorty," all aimed at driving up prices in the day-ahead and the real-time markets in one way or another.

It was arguably one of the most complicated heists ever undertaken in California. Very few people understood what was going on, but it wasn't difficult to see that someone, somewhere, was making out like a bandit. In October 2000, a man who lived near San Diego sent an email to Enron's public relations department congratulating the company on its quarterly earnings. He mentioned he had paid more than $350 for electricity during the summer months. "You cannot sit there and tell me that you did not benefit from the exploding energy prices caused by 'DEREGULATION,'" he wrote. "Keep sucking up those profits. . . . Like a vampire."

That fall, the utilities were running out of money to purchase power. The crisis quickly got political. The governor who had championed deregulation had been replaced in 1999 by Gray Davis, the first Democrat in sixteen years to hold the office. Two years into his term, it became clear that any reelection campaign could very well live or die by how he handled the disaster. He was loath to raise retail electricity rates, which to him seemed akin to political suicide. His challenge was to arrive at a solution that would keep the lights on without forcing utility customers to swallow the actual cost of power.

For Glynn, time was of the essence. PG&E was racking up billions of dollars in debt, and its shareholders were breathing down his neck. He had brought on bankruptcy advisers as early as September as a precautionary measure. The head of the CPUC had deferred to the governor on whether the state should step in to help the utilities. Whether PG&E resorted to Chapter 11 would depend on whether it could strike a deal with Davis.

For that, Glynn needed Dan Richard, PG&E's senior vice president for public policy and government relations. Richard, tall and broad-shouldered, had been with PG&E for only a short time, but he was totally committed to the job, intent on protecting the company from what threatened to be an unmitigated financial disaster. He had once been described as having a "warrior mentality"—he would keep fighting even after he was totally spent. And the fight was only just beginning.

Richard was as sharp as he was passionate. He understood the complexity of the problem PG&E faced. He had spent years as an adviser to power companies looking to build plants in California and had been involved in shaping the deregulation bill. He knew most everyone in Sacramento. He made some calls, trying to set up a meeting with the governor. But Davis kept his distance as his advisers debated what he ought to do.

In December, as the grid operator narrowly managed to avoid more rolling blackouts, Davis finally agreed to a meeting with the heads of the three utilities. Glynn and Richard arrived in Sacramento to sit down with Davis, the president of the CPUC, and the leaders of the senate and the assembly. PG&E and Southern California Edison dominated the discussion. The two utilities were together in an $8 billion hole. Both wanted to unfreeze retail rates and begin charging customers as much as 30 percent more for power. Davis wouldn't agree. He might have allowed for more modest rate increases if the utilities agreed to shoulder some of the costs themselves, but the representatives for PG&E and Southern California Edison all argued it wouldn't be enough. The meeting ended in a stalemate.

Senate president John Burton, a fiery Democrat from San Francisco with an irreverent streak, got up to leave. Before he reached the door, he turned and tossed a crumpled dollar bill on the table to remind the governor that the crisis had become his to fix—a stalemate couldn't last without major consequences for both the utilities and the state.

"What's that?" Davis asked.

"The buck, Governor," Burton replied. "It stops with you."

"I'm pretty sure it landed in front of PG&E," Davis replied.

There was another wrinkle in California's effort to stop the utilities from bleeding out. Most any deal needed the support of the Federal Energy Regulatory Commission, which had assumed oversight of wholesale power markets during deregulation. But FERC was hesitant to take a stand. Momentum toward deregulation had started with the federal government, which had allowed states to test different market designs. Regulators feared that intervening in the California crisis would tar the sanctity of the states' free-market experiments. And they were caught between two administrations as President Bill Clinton handed the reins to George W. Bush, who, as governor of Texas, had supported a sweeping deregulation effort there.

Around the time Davis met with the utilities, he asked FERC to impose a cap on wholesale prices, which had at times topped $1,400 a megawatt hour that month. The regulators, after much waffling, decided on a "soft cap" of $150 per megawatt hour, but included a provision to allow suppliers to charge more than that. The decision was effectively useless. US senator Dianne Feinstein, incensed by the crisis in her home state, accused the regulators of fiddling while Rome burned.

The incident evoked the irony of a scene in *Attack of the Killer Tomatoes!* in which the large, plastic-looking fruits roll through the streets of idyllic California towns, laughing maniacally in search of their next victims. The federal government can't manage to stop them. The president spends his days signing meaningless pieces of paper, each one with a new pen. He directs his chief of staff to hire a public relations man to convince terrified citizens that things could be worse.

With a flourish, the public relations man unveils an easel outlining his campaign, titled "Tomato Plants vs. Nuclear Plants." Among the items: tomato plants cost less to build.

"Who can argue with that?" the man booms.

Come January, Glynn had nearly reached the end of his rope. Richard drafted a strongly worded letter to Davis on Glynn's behalf, imploring him to use the state's purchasing power to either buy electricity or guarantee PG&E's tattered credit. State leadership, Richard wrote, had months to deal with the crisis and yet still hadn't done so.

"Many of the options available to the state have been lost in useless squabbling with federal agencies," he wrote. "The company is collecting about $1 million per hour less than it can recover in rates. We have warned that this situation cannot continue."

The crisis came to a head just a couple of weeks later, when the grid operator faced a serious power shortage and sounded the alarm.

A number of power plants had gone off-line, ostensibly for maintenance. Enron saw opportunity to drive up prices by surreptitiously taking another off-line. On January 16, as the grid operator pleaded for Californians to conserve power, an Enron trader identified only as Bill called up a coconspirator named Rich, who oversaw a small power plant in Las Vegas.

"Hey, Rich, this is Bill up at Enron," he said.

"Bill Junior or Senior?"

"The third."

Bill III then asked Rich if he could take the power plant off-line the following day, just before the afternoon surge in demand. "We want you guys to get a little creative and come up with a reason to go down."

Rich replied that he could idle the plant under the guise of inspecting a switch on a steam turbine. The electrician on staff could check it that evening, and find cause to check it again the next day. The plant would suddenly go down for maintenance at a time when it was supposed to supply power, forcing the grid operator to pay huge amounts of money to replace its output.

"I think that's a good plan, Rich," Bill said. "I knew I could count on you."

The following day, as the electrician supposedly checked the steam turbine switch, the grid operator again pleaded for utility customers to conserve electricity. It wasn't enough to avert a crisis. Just before noon, the grid operator ordered PG&E to cut power to half a million households and businesses in parts of Northern and central California.

The blackouts continued the next day. Late that afternoon, as traffic snarled and the sun set on darkened buildings, the Power Exchange suspended trading privileges for PG&E and Southern California Edison, which had defaulted on hundreds of millions of dollars in power purchases and seen their credit ratings reduced to junk. That night, Davis declared a state of emergency and authorized the California Department of Water Resources to begin purchasing power on behalf of the utilities.

Sladoje felt beaten down and hopeless. The Power Exchange had indeed made history—its name would become synonymous with the crisis. At a meeting one Sunday in January, he told the exchange's board that there was no use in going on. The exchange folded at the end of the month and filed for bankruptcy protection in March. In a statement before the court, Sladoje asked for permission to keep paying a handful of staffers as they attempted to collect overdue payments from PG&E and Southern California Edison. The spot power market was dead.

By then, the state had spent billions of dollars buying power on behalf of the utilities, which were still drowning in debt. Richard was working nights and weekends, debating various measures with the governor's office. The company eventually agreed to sell its transmission system to the state in a desperate effort to get the cash it needed. But the parties couldn't agree on the terms of a sale. Richard haggled with Davis's team for weeks. By the end of March, negotiations had stalled. Glynn told Richard that he had lost faith that a deal would ever materialize.

Negotiations picked back up, only to fall apart. On April 3, a Tuesday, Glynn decided that Pacific Gas and Electric Co., the utility subsidiary of the new PG&E Corp., would seek bankruptcy protection. The company's lawyers sprang to action. They planned to file the papers that Friday.

PG&E would become one of the few utilities to have ever undertaken the long and arduous Chapter 11 restructuring process. Utilities rarely wind up in bankruptcy court because they rarely run out of money—state and federal regulators all but guarantee their ability to produce revenue and shareholder returns. A spate of them sought or neared bankruptcy protection in the 1980s, after overspending on nuclear plants. Since then, PG&E had been one of the only major ones to decide to do it.

Richard was working out of PG&E's office in Sacramento when Glynn made the call. It had been a long, difficult day. Despondency had set in. Night had fallen by the time he mustered the energy to pack up. On his way out the door, he ran into an old friend, a political organizer who was preparing to meet with Davis. The organizer mentioned that Davis was planning to give his first statewide address on the energy crisis on Thursday evening. Richard's stomach sank. He told his friend he thought that was a very bad idea. He just couldn't tell him why.

Richard and Glynn watched the speech together that Thursday. Davis, a thin man with sharp features and neat, sandy hair, addressed the people of California from the leather chair at his desk. He explained the roots of the crisis in the plainest of terms, speaking as simply as a children's television host. He condemned the "flawed deregulation scheme." He blasted FERC for its inaction. He touted his administration's efforts to ease the supply crunch by building more power plants and promised to go after those who were bilking the system. Then, he surrendered to raising retail electricity rates. He said that many residents would see their bills rise by nearly 27 percent. Large businesses would face even steeper increases.

"You know I've fought tooth and nail against raising rates," he said. "It's become increasingly clear, however . . . that some increases are necessary to keep our lights on, and our economy strong."

Richard looked at Glynn. "Does this change anything?" he asked.

Glynn shook his head. In his view, it was too little, too late. The train had left the station. The lawyers had drafted the papers, and the communications team was bracing for the day to come.

The next morning, PG&E filed for bankruptcy protection. It hosted a press call. More than eight hundred people dialed in.

"The kindest thing to say is that progress has dramatically slowed," Glynn said. "The regulatory and political processes have failed us, and now we are turning to the court."

For Davis, it was a slap in the face, given that he had just made a major concession to the utilities and tried to assure their creditors that negotiations were underway. Standing before a pool of reporters, he said that PG&E had "dishonored itself." The company's decision embarrassed California lawmakers and threw yet another wrench into their efforts to end the crisis. Whatever goodwill the company had left in Sacramento evaporated.

The Department of Water Resources issued bonds to relieve the $10 billion in debt it had incurred as a result of buying power on behalf of the state's utilities, money that had been advanced from the state's general fund. Utility customers would slowly pay off the bonds through a special charge tacked onto their bills, which were already set to become substantially heftier as a result of the crisis.

With that, California essentially halted its push toward deregulation. The CPUC suspended plans to allow customers to choose their electricity providers. That meant most everyone would keep buying power from the utilities, and the CPUC would resume its role in determining how much they could charge customers. The utilities, however, would never again play the same role in producing electricity as they had before the crisis. The wholesale power market would be resurrected under the watch of the grid operator. This time, the CPUC would allow the utilities to lock in prices with long-term electricity contracts.

The crisis was a lesson in what not to do in overhauling the provision of a critical service. In moving so quickly to restructure the entire industry, lawmakers and regulators failed to anticipate the sorts of contingencies that could seriously stress their peculiar market design. More than a dozen other states pumped the brakes on their own deregulation efforts, hesitant to unleash the forces of competition after watching them devour California. The result was a patchwork of wholesale markets across the United States, all with different designs and degrees of regulation. In most regions, utilities would remain the primary retailers, and regulators would continue to set their rates. Generators and traders, once seen as innovative and efficient, struggled to survive as the competitive power business went through a protracted slump. The largest among them was Enron, which was found to have been manipulating its balance sheet as well as the California power market.

Steve Peace, still a state senator, had tried his best to help capture the monster set loose by the deregulation bill. When prices first began skyrocketing, he called for the grid operator to cap wholesale power prices at $250 a megawatt hour, but the governing board narrowly voted against it in favor of a much higher ceiling. He tried, to no avail, to persuade the utilities to simply stop purchasing such expensive power. He framed a question to the president of Southern California Edison: If the cashier at Wendy's charged him $5,000 for a hamburger, would he agree to pay it? Answer: Of course not, but the utility had a duty to serve, and would leave it to the regulators to take action on prices.

As the crisis wore on, Peace became isolated, a pariah among his fellow lawmakers. There wasn't much he could do. Eventually, as he was advising a litigation effort targeting the market manipulators, he got a call from Davis, who was desperately seeking advice and needed someone who understood the problem.

It wasn't all Peace's fault. The bill he had stewarded was one of political compromises. Despite all the late-night negotiations and the line-by-line review, it had serious weaknesses that became most apparent in hindsight. Peace would later acknowledge a number of things that could have reduced the risk of disaster. He felt prices would have been better kept in check if the grid operator had a different governance structure, and wished more had been done to ensure accountability from federal regulators. But one of his biggest mistakes, he recalled, was underestimating the traders' ability to manipulate the market—and overestimating his own ability to take them on. Peace had spent his entire life—first as the small kid on the football field, then as an inexperienced movie producer, then as a young lawmaker—winning by virtue of being tough and clever. Up until that point, he had never gotten his ass kicked.

The crisis cast a pall over Peace's political career, one he later said he had all but lost interest in continuing. As he considered a run for secretary of state, his production company produced a video in which he attacked what he considered the "myths of deregulation," among them that he was an "architect." But the damage was done. "He couldn't be elected dogcatcher," one political consultant sniffed. He ended his run in Sacramento as state finance director under Davis.

PG&E, meanwhile, pieced itself back together in bankruptcy court, moved to get rid of its unregulated businesses, and debated who should pay its debts. After three years of negotiations, it reached an agreement with the CPUC. PG&E would reduce rates in the near term but collect from customers as much as $8 billion in extra revenue over the next decade.

Dan Richard, the government affairs head, and several other PG&E representatives appeared before the *San Francisco Chronicle*'s editorial board to convince a roomful of skeptics that the deal was a win all around. If PG&E collected less money from customers, Richard told them, the utility might have to resort to issuing high-interest junk bonds. That would also add to customer bills over time. The deal, as structured, was a bid to restore the company's investment-grade credit rating.

"The judge and jury is Wall Street," Richard said.

Five

TRANSFORMATION

P eter Darbee knew what made Wall Street investors tick. He had, after all, been one of them, rising to become a Goldman Sachs investment banker focused on energy and telecommunications. He parlayed that experience into executive roles at Pacific Bell and then Advanced Fibre Communications. PG&E hired him as its chief financial officer in 1999, just months before the energy crisis began.

Darbee stuck with PG&E through bankruptcy, helping Bob Glynn sort through the company's debts. It was exhausting, grueling work. Foam placards appeared around the building to keep employees motivated during the reorganization. One read STAY THE COURSE! with a weary-looking bulldog emerging into the sun. Glynn was determined to see the company through, but he had also started to consider his successor. It was only a matter of time before the board would look for a leader who could start fresh. Darbee seemed like a logical choice.

Across the country, it was becoming more common for CFOs to become CEOs. Board members saw value in having a financial mind to navigate increasingly globalized markets. And corporate accounting rules were getting stricter and more complex. The Sarbanes-Oxley Act, a federal law passed in 2002 in response to Enron's collapse and several other major accounting scandals, required more financial accountability from corporate executives.

Glynn, an engineer at heart, had tried his best to please the company's investors. He had worked hard, in the 1990s, to grow the shareholder base from a limited group of long-term investors to ones interested in quicker payouts expected to come from new competitive businesses. The bankruptcy had ironically served to expedite that, with hedge funds emerging to own more than a tenth of the company's stock for the first time in its history. Darbee looked like the sort of executive they'd like to see at the helm.

The utility exited bankruptcy in April 2004 with a strong balance sheet, its investment-grade credit rating restored. In December, PG&E announced that Darbee would replace Glynn at the start of 2005. Darbee, well groomed and well versed in the language of investors, vowed to turn PG&E into a modern utility, with real moneymaking potential.

That notion had traction on Wall Street. Long considered "widows and orphans" stocks—the stuff of retirement funds—utilities had, in the years prior to the deregulation push, generally been expected by investors to deliver little more than steady, modest returns. That changed when a Texas utility called TXU Corporation proved it could boost earnings and dividends well above historical levels.

Like PG&E before it, TXU had struggled to compete in a deregulated electricity market. Texas in 2002 became the first state to fully embrace deregulation by disintegrating the state's utilities, the result of legislation requiring them to separate their operations into distinct entities providing for the generation, transmission, and distribution of power. Texas Utilities, in preparation, had reorganized as TXU, with three subsidiaries managing power plants, wires, and retail accounts. It was a complex juggling act. The main problem, though, was the company's operations in Britain, which had charted a path to deregulation even before California, allowing foreign companies to go there and compete. TXU's British operations suffered as wholesale electricity prices fell, putting its long-term contracts out of the money. In 2002, the company lost more than $4 billion. Its stock plummeted. In 2004, on the edge of bankruptcy, it hired a shrewd executive named C. John Wilder to pull it back from the ledge.

Wilder, at forty-six, had earned respect as CFO at Entergy Corporation, a utility company serving parts of Louisiana, Arkansas, Mississippi, and Texas. He stepped in as TXU's CEO with an ambitious strategy that involved selling subsidiaries, outsourcing jobs, and buying back shares. He oversaw the sale of TXU's operations abroad and used the proceeds to pay down billions of dollars in debt. He further cut costs by modernizing work processes, outsourcing back-office functions, and cutting thousands of employees from the payroll, including dozens of executives who had been with the company for years. The savings went toward dividends and buybacks. By the end of 2004, TXU's share price had more than doubled to about thirty-two dollars a share, the highest in its history. It had tripled its dividend and repurchased more than 50 million shares. A stodgy utility had become a growth company, and Wilder became known as the "Turnaround Titan."

Darbee envisioned a similarly stunning turnaround for PG&E. He consulted with other executives to devise a sweeping overhaul of the way the company did business. Few could dispute there was ample room for improvement. PG&E had basic software systems, messy records, and budgeting problems. But like other utilities across the country, it hadn't found much incentive to take on the formidable challenge of redesigning processes that got the job done more often than not. Darbee wouldn't settle for that. He wanted PG&E to act more like a company that had to compete for customers and market share. The effort would be called "Business Transformation."

In February 2005, a month into the job, Darbee met investors at the St. Regis Hotel in New York City to explain his vision for the company. He would cut costs by modernizing PG&E's work processes and channel the savings into new power lines, pipelines, and customer programs. He estimated that the company had the potential to invest more than $9 billion in its gas and electric systems through 2009.

Utilities make money in an unusual way. Most companies generate revenues and subtract costs to arrive at a bottom-line number—profit, or lack thereof. Not so for utilities, which, as regulated monopolies, are almost guaranteed profits. State regulators set rates for gas and electric service that allow the utilities to recoup costs and make a reasonable return. There's another layer of complexity. Utilities earn returns for shareholders on capital investments, multimillion- or multibillion-dollar improvements that boost the overall value of their systems. Maintenance and operating expenses are treated differently. Regulators don't typically allow utilities to earn returns on the day-to-day projects and programs that keep pipelines and power lines running.

For those reasons, the industry's top financial performers are generally adroit at keeping expenses low and investing instead in projects that boost returns. That approach was core to PG&E's transformation effort. Darbee wanted to cut unnecessary costs and use the savings to make larger, more profitable investments.

Around the time he pitched investors on his vision, Darbee held nine all-day "transformation meetings" for PG&E's officers, managers, and supervisors. Thousands of employees gathered in cities from San Luis Obispo to Sacramento to hear him speak. His goal, he said, was to turn the company into the leading utility in the United States, with "delighted customers, energized employees, and rewarded shareholders." Employees snickered. They found the word *delighted* ironic, given the blackouts during the electricity crisis. Darbee didn't get the joke.

Two months later, PG&E restored dividend payments, which had been suspended during bankruptcy. Standing before shareholders at the company's annual meeting that April, Darbee said that the core business was finally back on track. He promised that the company would soon generate cash and then return it to them through dividends and buybacks.

"We know how important dividends are to you," he told them. "And I know I can speak for the entire company when I say that we're delighted to be paying you a dividend again."

In June, with the company's share price up 20 percent since its emergence from bankruptcy, Darbee appeared on CNBC during the lunch hour for a rapid-fire interview with anchor Becky Quick.

"Wall Street always wants to know, 'What have you done for me lately?'" she said. "So what are you doing next?"

"We're really focused on the question of taking care of our customers," Darbee replied. "And so we're transforming our business with a focus on trying to serve our customers faster, better, and more cost-effectively."

There was another reason for the transformation: the 2005 repeal of a Depression-era federal law that essentially limited utilities from expanding outside of certain geographic regions. The rollback of the law, called the Public Utilities Holding Company Act, opened the door for mergers and acquisitions that would allow the nation's investor-owned utilities to grow into multistate behemoths.

Warren Buffett's Berkshire Hathaway Inc. was among the first to seize the opportunity with the $5.1 billion acquisition of PacifiCorp, an electric utility serving 1.6 million customers in six northwestern states. The deal expanded Berkshire Hathaway's utility operations to serve 3 million electric and natural gas customers in ten states. Other big utilities were shopping, too, with money to spend.

Darbee wanted PG&E to follow suit and use its newfound financial might to grow by acquisition. In her interview, Quick asked Darbee what he made of the Berkshire deal, which had been announced just days earlier.

"That has everybody asking what's next," she said. "First of all, do you expect to see consolidation, and if we do, will you be someone who is buying or who is selling?"

"I do believe that we will see more consolidation in the energy business over time," Darbee replied. "And if I had to choose where PG&E will be in that game, we'll be a buyer, not a seller."

In the boardroom, Darbee hung a plaque titled "Expectations of Our Leaders." It outlined seventeen objectives. One said leaders should "protect both public and employee safety as the first order of business." Executives, however, were asked to sign a statement saying they understood that the company's number-one goal was transformation. The piece of paper made its rounds during an off-site meeting for company officers. Dan Richard, still in his government relations role, was among them.

"What about safety?" Richard asked as the paper made its way to him. Darbee and several other executives brushed off the question. More than anything else, signing was meant to signal solidarity. Richard passed it on without signing. His friends at the company later persuaded him to do so out of concern that Darbee would fire him.

One afternoon, one of Darbee's consultants appeared in Richard's office to make sure he understood the need to commit to transformation. Richard said he generally supported it, so long as it involved changing company management for the better. If it meant imposing changes on rank-and-file employees without requiring the same of executives, Richard said, the effort would fail. Employees would bristle at the hypocrisy. Richard pointed out something that had bothered him ever since he got an office within the C-suite on the thirty-second floor at 77 Beale Street in San Francisco's financial district. A set of elevators had been programmed to essentially eliminate any time an executive might have to wait to descend from his or her office to the parking garage. When someone on the thirty-second floor pushed the button, an elevator car would rise immediately, bypassing all the other floors as it went. Richard thought it was a ridiculous setup that signaled the collective arrogance of the company's top management.

Darbee's consultant listened and nodded. The clock hit 6:00 p.m. The men together walked out of Richard's office and stopped at the elevator bank. They chatted as Richard pressed the button. A moment later, the doors opened and a woman who thought she had been headed to another floor looked up in surprise to find herself in the executive suite. "Oh, I'm sorry!" she said, startled. Richard looked pointedly at the consultant.

The need to fall in line strained Darbee's relationship with PG&E's other executives and officers. He was a deeply unpopular leader. Many found him bombastic, controlling, and quick to anger. Meetings with him could turn good days sour.

Darbee was just as unimpressed with many of his subordinates, particularly some who had spent most of their careers at the company. He didn't feel they were up to competitive standards. He replaced as many as forty-two officers during his six years as CEO, often with hires from finance or telecommunications. The idea was to bring in fresh blood. But the turnover also meant a loss of institutional knowledge—the stuff that kept the company's gears turning.

By mid-2005, Business Transformation involved hundreds of employees from Accenture, the global business consulting firm. A cohort of newly minted MBAs marched into PG&E's headquarters. Employees eyed them warily. They gave the newcomers a snarky nickname: the Green Beans.

Accenture first tried to evaluate how PG&E's gas and electric operations stacked up to those of its peers. The consultants immediately encountered a problem. PG&E lacked records on the condition of its power lines and pipelines, making comparisons difficult. In a 2005 presentation to the company, Accenture stated the obvious: PG&E could better target its maintenance spending if it had better data. It suggested that the company invest in gathering it while also cutting costs.

Accenture proposed the deepest cuts to gas transmission and electric operations. The firm found that total spending in both areas had outpaced inflation during a ten-year period. It recommended slashing electric-system spending by as much as $500 million a year, acknowledging that regulatory requirements may "prevent the company from closing the gap." Accenture saw potential for further cuts by consolidating work processes, field offices, and control centers. It also pressed the company to outsource jobs and rely more heavily on contractors for pipeline construction and gas-leak detection. The consultants recommended axing as many as eight thousand people from the payroll, a number sure to invite the wrath of the company's labor unions.

Already, union leaders were concerned about the transformation effort. Chief among them was Tom Dalzell, who led negotiations on behalf of the company's most prominent union, the International Brotherhood of Electrical Workers Local 1245. Dalzell, always a little disheveled, had sandy hair and a distant gaze that gave him the look of a daydreamer. He was unorthodox in his approach to just about everything. He was an Ivy League graduate who had chosen the world of labor, working for Cesar Chavez and the United Farm Workers right out of college. He had the distinction of being one of the only people to pass the California bar exam without having gone to law school. And he had a sort of obsessive curiosity that he channeled into the study of American slang. He had made himself into an expert, with a room of floor-to-ceiling bookshelves holding tome after tome on linguistic oddities, many of which he had written himself.

All of this meant that Dalzell was not only beloved in his role but exceptionally good at it. He spoke the language of both white-collar executives and blue-collar workers, fearlessly questioning the former and fiercely protecting the latter. It didn't take long for him to see that the executives did not want the unions involved in designing the transformation. He saw what looked to him like a cultish commitment to the cause, and his members—the linemen, inspectors, clerks, and forepersons most affected by the changes—hadn't been asked to join in. He warned the executives that whatever Accenture had planned would never succeed without union input.

PG&E's rank-and-file workers had something that couldn't be digitized: tricks to make things work in a system with spotty records and software. Without their knowledge, any system overhaul would fail to account for all of the workarounds needed to keep things functioning. A prime example involved new software systems to reduce the wait for electric and gas hookups, which at the time averaged more than three months. Some employees, anticipating disaster, burned copies of the old software and hid them in their desk drawers. Their fears were founded. When the new process went live in late 2006, hookups came to a near standstill. Work orders disappeared to a place unknown within the new system. Managers eventually broke down and scrapped it. It was no way to sell the transformation gospel. As one consultant put it, it was the beginning of the end.

Meanwhile, that year, Darbee promised to increase earnings per share by at least 7.5 percent a year for the next five years, far outpacing most other utilities. And he raised the dividend by more than a third.

PG&E's internal struggles were obscured by another transformation: the transition to renewable energy. PG&E was becoming a global leader in buying solar and wind power and paving the way for other utilities to follow suit.

Arnold Schwarzenegger, the Austrian bodybuilder turned Republican politician, began charting a new course for California's utilities after he was elected governor in 2003. Voters had rallied to boot Gray Davis out of office in retribution for his response to the energy crisis. A raucous recall election saw the emergence of a slate of unlikely candidates, from television stars to political commentators. Schwarzenegger, an actor with no governing experience, was among them. He was quickly dubbed "The Governator," a nod to his starring role in *The Terminator* in 1984. He won in a landslide.

Schwarzenegger, who once demonstrated his environmental commitment by driving around in a hydrogen-powered Hummer, saw climate issues as the key to his legacy in California and across the country. Just weeks after taking office, he put in a call to Michael Peevey, the president of the California Public Utilities Commission. He invited him to Sacramento.

Peevey, an outspoken former utility executive with thick gray hair and rounded glasses, was well known within energy circles. He had left Southern California Edison during the deregulation debate to launch a retail electricity supplier that profited by capturing some of the utilities' biggest commercial customers. Peevey had also been instrumental in helping PG&E emerge from bankruptcy, helping to negotiate a settlement that satisfied the CPUC and gave the state more control over the lands around the company's hydroelectric network running through the Sierras.

His influence stemmed in part from his close ties to Gray Davis, the governor who struggled to address the energy crisis. The relationship ultimately resulted in his appointment to the CPUC. It started at the height of the meltdown, when Davis ran into Peevey at a Christmas party in Southern California. He pulled Peevey aside and asked him what he thought he ought to do. Peevey offered to write out some ideas. Davis gave him the number for a private fax machine in a back office of the capitol. Within a few days, the machine spit out a three-page memo with suggestions on how to reform the CPUC and use the state's resources to backstop the utilities. Peevey's phone rang a few weeks later, on a Sunday night in January. It was Davis's chief of staff. Davis wanted him in Sacramento. Peevey soon became one of the few people the former governor trusted for advice. After a meeting at the governor's mansion in the spring of 2001, Davis followed Peevey out of the room and pulled him aside. Davis asked if he would consider joining the CPUC. Peevey replied that he would—under one condition. He didn't want to be just one of five members.

"I'd have to be the president," he said. Davis agreed.

In many ways, Peevey was well suited for the job. He was a natural dealmaker, helped by early career stints with the US Department of Labor and the California Labor Federation. He had also cofounded several California advocacy organizations focused on economics and the environment. When he became CPUC president in 2003, he had two objectives: restoring the agency's reputation after the energy crisis, and pushing climate issues to the top of its agenda. He resolved to become the greenest president the commission ever had.

In Sacramento, after Schwarzenegger's election, Peevey and another commissioner took a seat in a tent in the courtyard of the governor's office. Schwarzenegger lit a cigar. He said he wanted to make California a global leader in reducing carbon emissions, and he wanted their help.

Schwarzenegger first zeroed in on rooftop solar power. Lawmakers tried and failed to enact a mandate that would require large homebuilders to offer their customers solar panels. So the governor turned to the CPUC. In 2006, the agency approved an initiative that provided nearly $3 billion in incentives for rooftop solar deployment over eleven years. That same year, Schwarzenegger signed the California Global Warming Solutions Act, which formally acknowledged climate change and pledged to reduce the state's emissions to 1990 levels by 2020, significantly below what they were expected to have been otherwise.

Achieving that goal meant that California's utilities would have to change the sort of power they purchased. The shift had already started with a state law passed in 2002 requiring them to supply a fifth of retail electricity sales with wind, solar, and other renewable sources by 2017. Legislation passed in 2006 accelerated that mandate by changing the deadline to 2010. The utilities would have to move much faster in order to comply.

At the time, the climate change debate was mostly limited to scientists and policymakers awakening to the consequences of burning fossil fuels with abandon. That changed with *An Inconvenient Truth,* a documentary by former vice president Al Gore that walked through the devastating effects of a warming planet. The film earned two Oscars and became a watercooler topic in corporate offices across the country.

Peevey got a feel for the movie's impact one evening at the Commonwealth Club in San Francisco. There, Peevey bumped into John Burton, the colorful state senator who had told Gray Davis that the buck stopped with him. Peevey asked Burton, who had since left the senate, if he had seen the film. He had.

"What'd you think?" Peevey asked.

"Let me give you the bottom line," Burton replied. "Mike, you and I are okay; our kids are probably okay; but our grandkids are fucked."

Other legislative types were just as receptive to Peevey's climate ambitions. He next looked to the utilities. Dan Richard and others on the governmental relations team recognized political opportunity in supporting the effort. Doing so would burnish PG&E's reputation and help it regain some of the goodwill it had lost during the bankruptcy. And it was a natural move for the company, which had long shown an interest in climate matters. It was an early supporter of state legislation to create a climate change action registry and had been one of the first utilities in the nation to begin formally measuring its carbon emissions. In February 2005, Richard wrote Darbee a memo with some thoughts on how the company should position itself on climate change.

"Our task is to respond favorably to our close-to-home regulators, and, I might add, to the beliefs of many of our customers," he wrote. "We enhance our prospects for good regulatory relations if our overall policies are viewed as progressive."

At the time, Darbee had been mulling potential acquisitions, including several utilities with a lot of coal-fired power plants. Richard advised against it. The company couldn't claim to be progressive if it kept adding dirty generators to its portfolio. Why risk another war with the state just a few years after the energy crisis?

Darbee understood. Though he had never considered himself much of a progressive, he had been thinking seriously about what climate change meant for a company like PG&E. He decided to explore the question exhaustively. Part of the process involved a series of events in which Darbee invited academics and thought leaders to share their perspectives on climate change. Elon Musk, then new to Tesla, appeared at one to talk about electric vehicles. Peevey, sometimes a speaker, sometimes in the audience, developed a certain respect for Darbee. The two men began spending more time together—environmental events, a San Francisco Giants game. Climate was always on the agenda.

In 2006, Darbee told shareholders that the company had an

urgent responsibility to source more renewable energy, which at the time supplied just a fraction of its generation. The company began signing wind and solar contracts. "If you had asked me five years ago, this wouldn't have occurred to me," Darbee told the *San Francisco Chronicle* that year. "Somewhere in this process, I developed a point of view."

The effort put PG&E and California itself in the national spotlight. Schwarzenegger appeared on the cover of *Newsweek* in April 2007, balancing a globe on his finger. "Save the Planet—or Else" read the headline. The following month, *Vanity Fair* photographed Darbee on a bluff overlooking the Golden Gate Bridge alongside activists, artists, and other climate change crusaders.

"It might seem strange to have the chief executive of a big utility in these pages, but here is Peter A. Darbee of San Francisco," the caption read. "Unlike others who sit at the controls of power plants, Darbee believes global warming is indeed a serious threat."

Between 2007 and 2010, PG&E signed contracts supporting the development of nearly three gigawatts of solar power. Southern California Edison contracted for 1.8 gigawatts over the same period of time. It was a historic amount. Elsewhere in the country, there was only half a gigawatt of solar, and 8 gigawatts globally. A gigawatt is a unit of power equivalent to 1,000 megawatts, requiring more than three million solar panels to produce. Sprawling solar farms would soon blanket California's Mojave Desert and Central Valley, with transmission lines stretching north to deliver the power to PG&E's customers.

It was enough to earn Darbee an invitation to appear before the United Nations at its 2008 Investor Summit on Climate Risk. On Valentine's Day, Darbee arrived at the UN headquarters in New

York City to explain the complexity of climate change and the need to set national targets for reducing emissions.

"The first thing to remember is that climate is a long-term challenge," he said. "If we're nearsighted in our thinking, we're going to make a lot of bad choices, both financially and environmentally, because we'll misunderstand the risks and opportunities—or miss them altogether."

As PG&E looked to the future, signs of climate change were already emerging in its home state. In October 2007, the winds were fierce in Southern California. These were the Santa Ana winds, devilish gusts that originate in the Great Basin, the vast system of aquifers, rivers, and lakes that hydrates much of the arid West. The winds occur each year during the fall and winter when pressure builds over Nevada and Utah, churning air up and over the mountains lining Southern California. The winds, hot and dry, pick up speed as they sweep down the western slopes to buffet Los Angeles and San Diego.

The winds themselves are nothing new. They have occurred for centuries, long before humans were around to feel them. But that fall, in 2007, they were uncommonly strong, whipping through the region with hurricane force. The fastest gusts topped eighty-five miles per hour. Southern California was also uncommonly dry that year, after a rainless winter and a punishingly hot summer made kindling out of shrubs and brush.

On the night of October 20, a wildfire broke out in the foothills north of Los Angeles. Two more ignited in the dark morning hours near Malibu and Santa Barbara. Another started at sunrise south of San Diego near the Mexico border. Cal Fire dispatched planes loaded with flame retardant.

One of the pilots was flying over the mountains northeast of San Diego when he saw a flash of blue—an arc of electricity jumping from a damaged power line owned by San Diego Gas & Electric. Sparks settled in the dry grass, igniting what became known as the Witch Fire. It spread with devastating speed. The fire burned nearly two hundred thousand acres, spreading west from the rural foothills to the San Diego suburbs. It killed two people, destroyed more than a thousand homes, and injured forty firefighters.

It was a harbinger of disaster to come. The fire's destruction marked the convergence of three distinct threats: broken power lines, devil winds, and climate change. Scientists were beginning to recognize that Southern California was getting warmer, with more heat waves and less rain. That change would stress the region's water supplies and increase the risk of wildfire, especially when the Santa Anas raced westward.

The fire had serious financial implications for San Diego Gas & Electric, which faced hundreds of millions of dollars in potential liability costs. At issue was a state constitutional provision known as "inverse condemnation." The concept sounds complex, but it's fairly straightforward. Under what's known as eminent domain, the government has the right to seize private property for public purposes, such as transportation or water provision, so long as it compensates the landowner. Inverse condemnation flips that idea by giving landowners a right to compensation if whatever is built to serve a public purpose somehow damages their property. The idea is to protect the landowner from contributing "more than his proper share to the public undertaking."

The California constitution was amended in 1879 to hold state and local governments accountable on the principle of inverse condemnation. Over time, that universe expanded to include municipal utilities such as the Los Angeles Department of Water and Power. Investor-owned utilities didn't fall squarely under that umbrella until 1999, when a California appellate court held Southern California Edison liable for property damage caused by a 1993 wildfire sparked by one of its power lines. The court determined that private utilities were substantially similar to their public counterparts. In essence, if a power line starts a fire, its owner is responsible for the damage, regardless of how the line was maintained. California is one of the only states where such strict rules apply.

San Diego Gas & Electric settled liability claims related to the Witch Fire and two others for $2.4 billion and asked regulators for permission to recoup a portion of those costs by passing them on to customers. It would take the CPUC years to arrive at a decision, with critical consequences for PG&E.

———

At PG&E's annual shareholder meeting in 2007, five union members stood up with a message that no investor wanted to hear: the gas division had serious problems. The men had tried for years to get the attention of company executives, but to no avail. So they showed up in a place where they couldn't be ignored.

Early on, Accenture had informed PG&E that it had a low frequency of gas leaks relative to other utilities. That turned out to be an illusion. The gas division was suffering as a result of a poorly designed incentive program that rewarded supervisors whose crews found the fewest numbers of leaks. Instead of finding and fixing more problems up front to reduce the number over time, they simply reported fewer of them from the start. To the workers, the problems seemed particularly acute within the company's network of distribution pipelines, which branch off larger transmission pipes and deliver gas to homes and businesses. The distribution pipes had been springing a dangerous number of leaks.

The transformation effort was one of several things that had obscured the problem by diverting the company's focus from the day-to-day obligations of running a utility. Executives overseeing gas and electric operations had found it increasingly difficult to argue for stepping up power line and pipeline inspections and maintenance, for that sort of spending had the potential to compromise the earnings growth Darbee had promised shareholders. The tension epitomized the challenge every investor-owned utility faces: finding the right balance between public and private interests.

After hearing the workers' concerns, Darbee called for an investigation. An internal audit exposed problems on a massive scale. Gas supervisors had for years been producing inaccurate leak surveys, which meant any number of leaks had gone unreported and unrepaired. The company sent crews out to complete an inspection blitz. Darbee met with the board of directors in March 2008 and gave its audit committee an overview of the findings. The problems were serious enough, he said, to bring in independent investigators to scrutinize gas distribution maintenance. The committee agreed.

It was too little, too late. Months later, on the morning of Christmas Eve, PG&E received a call from a woman in Rancho Cordova who smelled gas near her home on Paiute Way, a short street of small ranch-style houses just east of Sacramento. The company dispatched a representative, who arrived on scene just after 10:00 a.m. Residents pointed the representative to a neighbor's front yard, where an underground gas leak had choked out a patch of grass. The representative called for a leak investigator. The investigator got stuck in traffic. Then his truck had brake problems. It was early afternoon by the time he made it to Paiute Way. He spotted the brown patch and knocked on the door. A seventeen-year-old girl answered. She directed him to talk to her grandfather, who was out in the garage.

In the garage, the grandfather told the investigator that PG&E had been there once before, a couple of years earlier, to repair a leak in the same place. The pipe itself was made of crack-prone plastic called Aldyl-A, and PG&E had done a poor job repairing it. The investigator walked into the yard and placed a gas detector above the dead grass. It registered sixty thousand parts per million, a combustible concentration. Moments later, at 1:36 p.m., the granddaughter sneaked into the bathroom to light a cigarette. The house exploded. The grandfather died. Everyone else—daughter, granddaughter, investigator—was rushed to the hospital.

As PG&E grappled with its gas problems, Business Transformation had become its own disaster. The effort, snarled and ineffective, simply hadn't produced the savings that Darbee had planned to reinvest in the company. PG&E announced its earnings would fall short of analysts' expectations in the fourth quarter of 2007 and said it expected future savings to be as much as $285 million less than forecast.

Workers, meanwhile, were fuming. Almost everything had broken down during transformation. Accenture had attempted to centralize dispatch centers but failed to outfit them with the necessary technology, forcing dispatchers to rely on MapQuest and Google Earth to find where crews were supposed to be performing work. A new design tool meant to make work easier didn't jibe with PG&E's system. An effort to improve the way the company used its truck fleet created more problems than it solved. Even payroll had gotten bungled, making the workers angrier still.

One summer day in 2008, Tom Dalzell, who had an impish sense of humor, marched into the C-suite carrying a thick mock-up of the union's quarterly newsletter, which circulated among some twenty-three thousand employees and contractors. "PG&E's Billion Dollar Bust" was splashed across the top. "Far from delighting its customers and energizing its employees, PG&E's billion-dollar Transformation neglected the very things that matter most and have transformed PG&E into a 'Utility in Trouble,'" it read.

The newsletter contained a report card that graded each of Accenture's initiatives. Hardly any scored higher than a D, and there were more than a few Fs. It then went on to spotlight PG&E's deteriorating infrastructure, with pictures of problem-ridden poles and wires strung throughout Northern California. "It's not hard to find evidence that PG&E is a utility in trouble," it read. "Just look up."

The executives sighed. They asked what it would take for Dalzell not to print it. Stop sidelining the unions, he told them, repeating what he had said years earlier. The workers were best positioned to help fix the problems transformation had wrought.

It wasn't long before PG&E scrapped the transformation effort entirely. It had improved a few things, such as the way the company sourced materials. But it had done almost nothing to address problems with the company's aging power lines and pipelines, poor inspection practices, and flawed records, many of which were still kept on paper. Accenture charged PG&E more than $300 million for its services, but the company was so dissatisfied that it negotiated a roughly $30 million reduction.

"The initiative was too theoretical to implement in a practical way," PG&E said in a report to regulators.

The next summer, in 2010, the company again lost goodwill in Sacramento by funding a ballot initiative that aimed to make it harder for local governments to form electricity-buying authorities called community choice aggregators. Such authorities enabled a city or county to step into the role of the utility by generating or purchasing electricity, allowing it to set rates and buy more renewable energy for its customers. Clunky name aside, community choice aggregation was gaining traction in California, and Darbee was worried it would take over the utility business. For someone who wanted PG&E to act as though its customers weren't captive, Darbee wasn't keen on giving them an option to leave.

The rebuke was swift. The ballot measure had very little support within the political establishment. None of the other big utilities had backed it. Environmentalists opposed it. Voters ultimately killed it. If climate change showed Darbee at his best, Peevey later recalled, the $46 million lobbying effort showed him at his worst.

Investors failed to see the annual 7.5 percent increases in earnings per share Darbee had promised from 2006 to 2010. PG&E delivered an average of 3.6 percent during that time. It had, however, succeeded in earning its rate of return on equity, the profit margin authorized by regulators, by making large capital investments. Darbee later told *The Wall Street Journal* that the attempt to modernize the company was a "bumpy and difficult road," like "trying to change the direction of a huge, oceangoing tanker."

His failure to turn the ship came at a cost. Morale had fizzled among executives who had tried to champion the transformation, only to find they had wasted their time. And workers, more so than ever, were wary of change. The word *transformation* became taboo.

It had been a lost decade. But the worst had yet to come.

Six

SAN BRUNO

A t dinnertime on September 9, 2010, a gas transmission pipeline underneath the San Francisco suburb of San Bruno
turned into a firebomb. A sudden increase in pressure caused
the pipe to rupture at the seam. Gas seeped out and ignited within seconds. A three-thousand-pound piece of steel ejected from the earth,
leaving a crater nearly half the width of a football field. The projectile
crashed some one hundred feet from the crater's edge, right in the
middle of a neighborhood.

The 911 calls poured in seconds after the eruption. Local officials
were inundated with reports that a plane had crashed or a gas station had exploded. A news helicopter captured footage of the crater
spewing fire like a flamethrower. The fire mushroomed as it fed on
gas, towering sixty feet above the surrounding rooftops.

Within PG&E, terror and confusion spread as employees heard
the news. Those at the company's gas control center soon realized

that a pipeline had ruptured, and began phoning the dispatch center. Their calls were met with skepticism. News reporters were speculating that a small plane must have dropped out of the sky.

"It's easy to believe it's a plane crash," one control center operator said to a dispatcher. "We still have indication that it is a gas line break. We're staying with that."

The severed pipeline, large enough for an adult to crawl through, bled gas for more than an hour and a half as PG&E crews tried to figure out how to stop it. PG&E estimated that nearly forty-eight million cubic feet of gas—enough to fill more than five hundred Olympic swimming pools—escaped and burned. The fire destroyed thirty-eight homes and damaged seventy. Eight people died and nearly sixty were injured, with at least fifteen rushed to nearby hospitals. Four were sent to a burn center in San Francisco to start a long and painful recovery process.

Two days later, hundreds of people crowded into St. Robert's Catholic Church, just a couple of miles from the blast site. Shaken and angry, they demanded an explanation from PG&E. Peter Darbee did not appear. Instead, a relatively new executive named Geisha Williams, clad in khakis and a brown sweater set, took a microphone in the center aisle.

It was Williams's first big test in the California spotlight. She had spent most of her career at Florida Power & Light, where she had first interned as a young engineering student at the University of Miami. She returned after graduating and rose through the ranks of its electric division. She had been recruited to PG&E in 2007 to oversee the company's delivery of gas and electricity. She was hardworking and tough, but also kind and empathetic, with a finely tuned sense of how others might perceive her. PG&E sent her to St. Robert's as its chosen diplomat.

As sunlight streamed through stained glass, residents asked PG&E for the locations of other gas transmission pipelines. PG&E couldn't share them, Williams said, because doing so posed a security risk. The crowd hissed.

"That's bullshit!" someone yelled from the back.

Williams stayed calm. PG&E didn't know why the pipeline had exploded, she said. But it would find out.

"We know our customers are extremely nervous, and who wouldn't be?" she said. "We're going to be as responsible as we possibly can."

Investigators from the National Transportation Safety Board began sifting through the rubble. Within a month, they arrived at a preliminary conclusion. The pipe, laid in 1956, had been poorly welded. That made it vulnerable to the effects of sudden changes in gas pressure. Too high, and it could burst at the seam.

The findings raised serious questions about whether other PG&E pipelines had similar problems. State regulators were aghast. Paul Clanon, the executive director of the California Public Utilities Commission, typed up a concerned email days before the NTSB report went public. The recipient: Brian Cherry, PG&E's head of regulatory relations. Clanon asked whether PG&E could assure the public that its system was safe.

"Do your people actually have the data?" he asked. "Or should PG&E be doing an all-hands effort to make absolutely sure it knows what's down there for every pipe segment?"

"God knows what is underground," Cherry replied. "That said, they are working feverishly to come up with answers."

In January, as the investigation continued, the NTSB sent PG&E a letter with urgent safety recommendations. In its forty-four-year history, the agency had rarely found such cause for alarm. The company had listed the manufacturer of the ruptured pipe segment as "NA," meaning it didn't have the information, and had used flawed records in noting its fabrication material. The NTSB suspected it wasn't a one-off discrepancy. It recommended that the company sift through all of its gas records—millions of pieces of paper—to find information gaps and determine whether other pipelines were at risk of rupturing. The CPUC ordered the company to get started at once.

It was a gargantuan task. PG&E's records were an utter mess, stored in filing boxes all throughout its many offices in Northern California. Employees would have to centralize them and sort through them by hand. In early March, the company took over Cow Palace, a 250,000-square-foot arena south of San Francisco. Forklifts delivered hundreds of pallets stacked with tens of thousands of boxes.

As PG&E dusted off its aging files, Nick Stavropoulos got a call from a headhunter at his home in Boston. PG&E needed a complete overhaul of its gas operations, and it was looking for someone who could lead it. Whoever took the job could become a hero. But success would come at a cost. It would mean undoing years of mismanagement, facing down public vitriol, and figuring out what, exactly, was wrong with a pipeline system that had been deteriorating underground for decades. There would be almost no time off. Would Stavropoulos fly to California for an interview?

Stavropoulos knew old pipelines. He had started his career as an intern at Colonial Gas Company, a Boston-area gas utility with roots predating the Civil War. After several acquisitions, Colonial became part of KeySpan Corporation, one of the nation's largest gas distribution companies at the time. Stavropoulos rose to oversee KeySpan's operations in New England, New York City, and Long Island.

A crisis occurred on November 9, 2005, when a KeySpan crew set to work on fixing part of its system in the heart of Lexington, Massachusetts. The crew accidentally allowed high-pressure gas to flow into a pipeline feeding a home on a wooded cul-de-sac just north of downtown. An eighty-six-year-old distribution pipe, corroded and brittle, burst open. Gas seeped into the basement, around the furnace and the water heater pilot. The house, empty at the time, burst into flames.

KeySpan's attorneys advised Stavropoulos to avoid admitting the company was at fault. An investigation was underway, and the company was innocent until proven guilty. Stavropoulos disagreed. The company had obviously made a series of mistakes. Public trust had gone the way of the flaming house. Standing before a group of angry residents a day after the explosion, Stavropoulos said that the crew had made serious errors and that the company would take every measure to figure out what exactly had gone wrong.

"I take personal responsibility for what happened," Stavropoulos said.

The admission earned him respect within the community as well as the company. The following year, KeySpan was acquired by National Grid, an energy company based in England. Stavropoulos became its chief operating officer of US gas distribution, overseeing a complex network of pipelines spanning Boston, New York City, and Rhode Island. The early morning phone calls and flights back and forth to headquarters were tiring. Stavropoulos had been considering moving to another company. Then the phone rang.

PG&E booked Stavropoulos a room at the Hyatt Regency, about a block from the company's thirty-four-story headquarters in down-town San Francisco. He met with Darbee and other executives, sometimes in group interviews where everyone gathered around a conference table to talk about the gas system and what Stavropou-los could do about it. In his unmistakable Boston accent, he made one thing clear. He wouldn't stand for bullshit. If he was going to take the job, they needed to be straight with him about the extent of the problems.

He arrived again on April 21, 2011, for another round of inter-views. No sooner had he walked in the door than the president of the utility pulled him aside. The executive wanted Stavropoulos to hear it from him first: Darbee was out. The company was planning to make the announcement within hours.

Amid calls for his ouster, Darbee had gone to the board chairman and offered to resign within a few months. At that point, Darbee thought retirement was in his best interest. The chairman thought Darbee should move up his timeline, and he agreed. Darbee departed on April 30, with a $34.8 million retirement package.

Stavropoulos, meanwhile, was asked to head the gas division. He moved in June to a small downtown apartment walking distance from PG&E headquarters. Stavropoulos hit the C-suite like a bolt of lightning. He was pure energy, enthusiastic and quick to laugh. In a room full of black and navy suits, he might have been wearing a bright patterned jacket. But he also had a serious side, with a temper he worked hard to keep in check. He became a regular at SoulCycle in downtown San Francisco, pedaling out his frustrations to a fast beat in a dark room.

A few months into the job, Stavropoulos put on a pair of jeans and drove up to Vacaville, a small town roughly halfway between San Francisco and Sacramento. It was home to IBEW Local 1245's headquarters. Stavropoulos had set up a meeting with Tom Dalzell, the union leader, and about two dozen union members. He arrived alone.

The workers were agitated. One after another, a series of disasters had eroded their morale. The company had slashed head count in preparation for deregulation. More cuts came during the botched transformation effort. The workers had felt mostly ignored in the run-up to the Rancho Cordova explosion. And now, one of the worst utility disasters in American history had made it difficult for them to show their faces among friends and neighbors.

Already, many of the gas workers felt like second-class citizens within the company. As they put it, PG&E had always been a "big *E*, little *g*" utility. The sprawling electric system had historically gotten the most attention because it generated roughly three times as much revenue as the gas system and served a wider area. The gas workforce had shrunk to about half the size it had been twenty-five years earlier, prior to deregulation. Many of those who remained were approaching retirement age. If Stavropoulos didn't start hiring, the division would soon be short employees as well as institutional knowledge.

Stavropoulos understood the men on a personal level. He had blue-collar roots in Cambridge, Massachusetts. His father had worked in the meatpacking industry. His mother, a factory worker, had been an IBEW member for forty-one years. They shared a modest three-story home with other relatives. Hard work was expected. Stavropoulos got his first job at twelve.

He also understood tragedy. A single crushing loss a few years earlier had left him learning to cope. It had been a meal at home like any other. Then his twenty-six-year-old daughter choked on a piece of meat. Stavropoulos rushed over to perform the Heimlich. His wife called 911. But his daughter couldn't be saved. After her death, he found himself looking for a new sense of purpose. It was a big reason why he took the job at PG&E.

In the union hall, Stavropoulos sat next to Dalzell at the head of a large gray table. Stavropoulos wore a simple striped button-up, his thick brown hair a bit unkempt. Many of the men around the table wore baseball caps and fishing shirts and clutched paper coffee cups. When Stavropoulos spoke, he sounded like them—direct and unfiltered. He said he understood that past management had failed them. Then he told them what he wanted to do: overhaul the company's inspection and maintenance programs, known as integrity management, to identify all risks, no matter how much work it took.

"If we have to do a thousand inspections, we're going to do a thousand inspections," he said.

The workers were skeptical. They peppered him with questions. The Green Beans never would have green-lighted a thousand inspections for fear of what it would cost. Past executives might not have, either, for fear of what it would find.

"Words are cheap," Stavropoulos eventually said. "You've got to judge me by my actions."

Then Stavropoulos went quiet as the workers explained the extent of the mess that Darbee had left. Stavropoulos had heard about the transformation effort, but he hadn't realized just how little the company had to show for it. It wasn't as if the company had paid for a Jaguar and gotten a Camry, Stavropoulos would later say—"It paid for a Jaguar and got no car."

The workers were particularly concerned about Aldyl-A, the type of plastic pipe involved in the Rancho Cordova explosion. It was a DuPont product, manufactured in the 1960s and 1970s as an inexpensive alternative to steel or cast iron. Gas utilities across the country

began installing it throughout their systems. It soon proved dangerously prone to leaks.

Stavropoulos had spent a lot of time purging Aldyl-A from National Grid's system. He would do the same at PG&E. In the mean-
time, he said, crews needed to respond much more quickly when
customers reported smelling gas. It had taken hours for workers to
address the leak at Rancho Cordova. Then the house exploded.

Several union members offered ideas to fix the problems. Dalzell
watched as Stavropoulos took notes. "What we're seeing here is the
opposite of Accenture," Dalzell told the group.

"I don't hire consultants," Stavropoulos said. "They pay me to be
the manager, why hire somebody else with less experience?"

The men burst into applause. As he looked around the room,
Stavropoulos recalled a bit of free advice Dalzell had offered him a
couple of months earlier. Over breakfast at the Hyatt, Dalzell had
warned him to never use the word *transformation*. No doubt the gas
division needed one, but it would have to be called something else.

Peter Darbee's successor was just his opposite. More conservative
than transformative, utility veteran Anthony F. Earley Jr. didn't sweep
in with a new vision for PG&E when he arrived in September 2011. He
just wanted to get the train back on the rails.

Tony Earley learned his leadership style in the navy. As a young
officer on the USS *Hawkbill*, a nuclear submarine, he realized his
crews knew more than he did about steering the ship. His job, he
learned, was to remove the obstacles in their way. He called it ser-
vant leadership. After the navy, Earley went to law school and then
joined the utility world. He spent much of his career at DTE Energy

Co., which operates electric and gas utilities serving more than 3 million customers in Michigan. He became its CEO in 1998. Like other utility executives at the time, he had spent the early 2000s focused on acquisitions and investments in technology. And Michigan, like other states around it, had begun to deregulate its electricity market. He attempted to prepare.

Then, a career-defining crisis occurred on the afternoon of August 14, 2003. Earley was sitting in his office on the twenty-fourth floor of DTE's high-rise in downtown Detroit when the lights flickered, then died. His phone rang. For the first time in history, the company's entire system had crashed. All of its customers were in the dark. Earley felt his way down the stairwell and emerged into the parking lot, where a pool of reporters had gathered in the summer heat. Tie loose and hair a mess, he launched into the first of seven press conferences he would hold in the next three days. It wasn't just DTE, he told them. The outage had swept from New York to Canada, stranding travelers, cutting off emergency services and leaving millions sweltering in their homes. The cause was unknown at the time. But it was clear that the problem had disrupted the balance of supply and demand on the grid. The result was the sort of cascading failure that the California grid operator had narrowly avoided by resorting to rolling blackouts a few years earlier.

"This is an operation that will take several days to get the system back to where it's totally normal," Earley said. "It's a very complex process."

Earley worked almost around the clock for the next several days, subsisting on Fritos and yogurt until power was restored the morning of August 16. Investigators traced the root of the problem to First-Energy Corp., an Ohio company with several regional utilities. Trees had brushed against several of the company's transmission lines, causing them to trip off. The sudden disruption threw off the grid's calibration, causing power plants across the region to shut down in a matter of minutes. The outages affected about 55 million people.

The blackouts spurred the federal government to require better coordination among power producers, utilities, and regional grid operators. It expanded the oversight role of the Federal Energy Regulatory Commission and gave it charge over the North American Electric Reliability Corporation, a longstanding industry oversight body that had little enforcement power. After the blackouts, the feds gave NERC some teeth. The organization became responsible for setting stricter reliability standards for the utility industry. Transmission lines became a key focus. NERC drafted new requirements to ensure utilities kept their lines in good condition and clear of trees.

Michigan leaders lauded Earley for staying calm and clear-eyed throughout the crisis. He maintained that respect by keeping the utility running well in the coming years. It had only taken a few errant branches to bring down a big part of the grid. Earley invested in making sure that DTE wouldn't find itself in the same position as FirstEnergy. He was approaching retirement age when PG&E tracked him down. His instinct was to steer clear, but he reconsidered. "My first reaction was, 'Why would I want to do that?'" Earley later recalled. "My second reaction was, 'I think I know what needs to be done. I think I have some of the skills.'"

Earley joined PG&E in September 2011. At sixty-two, he still had thick dark hair and a boyish face that had softened a bit with age. Two weeks after his arrival, the NTSB released its final report on San Bruno. The 153-page document spelled out, in painful detail, the company's failure to maintain the safety of its gas system. To start, PG&E had few records on the pipe segment that ruptured. Those it did have proved to be inaccurate.

The pipe in question had been laid in 1956 using a hodgepodge of material left over from other projects. PG&E didn't know when or where it had been fabricated. A company database recorded it as a seamless pipe segment. In fact, it did have a seam, and a poorly welded one at that. The shoddy welds ruptured under pressure. Investigators found that the segment had been defective from the start, and would not have met quality control or welding standards even at the time it was installed. But PG&E hadn't detected the problem. The pipe sat, like a ticking time bomb, for fifty-four years.

Earley soon announced a strategy that he called "back to basics." He promised to restore public trust in PG&E and committed to spending an extra $400 million over two years to accelerate repairs to the company's gas and electric systems. Those costs would be borne by shareholders, not customers. Earley told investors to expect lower profits the following year.

Longer term, however, customers would help pay for the costs of overhauling the gas network. Shortly after Earley arrived, the company unveiled a plan to spend more than $2 billion to pressure-test 783 miles of transmission pipeline and replace large sections of the system. The CPUC authorized the company to raise rates to recoup at least half that amount from customers.

"We need to be brutally honest about where we are," Earley said in early December, three months after he became CEO. "Right now, we need to be humble because we have a lot to be humble about." He issued a warning to employees: "We cannot afford to have any significant operational issues."

A week later, a PG&E power line plunged a nationally televised *Monday Night Football* game into darkness. Earley himself was there. He had just walked into the owners' box when the lights cut out. A wire powering the stadium had fallen to the ground. The San Francisco 49ers and the Pittsburgh Steelers could barely see each other on the field. The floodlights surged back on. Ninety minutes later, there was a second blackout. Brian Cherry, the regulatory relations head, fired off an email to a utility commissioner lamenting the company's bad luck at such a conspicuous moment.

"We are snakebit," he wrote.

"LOL!" the commissioner responded. "Give Tony my best."

Weeks after Stavropoulos arrived, he sent crews into the field to inspect and pressure-test hundreds of miles of pipeline. The testing required draining the lines of gas and forcing pressurized water through the system. If a pipeline burst, water would spew out. In October 2011, crews testing a pipeline outside of Bakersfield heard a blast. The pipeline ruptured at a seam, blowing a crater in an alfalfa field. It was one of three failures Stavropoulos encountered that year as the company tested 160 miles of pipelines. All were repaired and retested.

In 2012, the company brought in Lloyd's Register, the venerable British risk-management consultant, to evaluate its gas operations. It found that the utility had fallen far behind peers in modernizing pipeline records and using state-of-the-art methods to monitor operations. The company had been using mostly the same equipment and processes for decades, even as new technology made patrols and inspections easier and more effective.

The pipelines were simply not up to modern standards. Across the country, utilities were moving to upgrade their gas systems so that they could be monitored with tools known as "smart pigs," digital devices that swim through pipelines and look for problems using

GPS mapping data, magnetic sensors, and other information. At that point, only about 15 percent of PG&E's system was equipped with the sort of technology the pigs needed to work.

One day, Stavropoulos climbed into a gas truck. He was struck by what was missing. There were no screens. Workers couldn't easily call up electronic records while they were out in the field. His charges at National Grid had been able to do that since the mid-1990s. So he upgraded PG&E's truck fleet. He then assessed the company's scattered gas control centers, with their antiquated paper maps and filing systems. He decided to bring them under one roof in a massive new facility in San Ramon, in the East Bay.

By the end of 2013, the San Ramon control center was complete. A ninety-foot wall of screens spanned one side of the control room so that staff could monitor 6,750 miles of transmission pipeline and 42,000 miles of distribution pipeline in real time. Computers with multiple monitors were spread throughout the room. There was also a smart board where employees could see what records field crews were using and mark them up as they worked.

It was an open floor plan. No one, not even Stavropoulos, had an office. He would walk around the control floor, joking with employees and working alongside them. And he had been hiring at a rapid clip, onboarding some two thousand people to work in the control center or out in the field. The crusty old gas division suddenly felt like a startup.

PG&E traditionally had spent five dollars to maintain its electric system for every dollar it spent maintaining its gas network. After San Bruno, maintenance spending skewed toward gas. In 2011, the company spent nearly $240 million to maintain its gas network, more than twice the amount in 2010. The numbers kept rising, peaking in 2016 at $658 million. Gas-system spending topped electric for the first time that year—briefly, PG&E became a "big *G*, little *e*" company. And it was spending billions more to make major capital improvements.

Lloyd's returned in 2014 to see how far PG&E had come. The consultancy gave the company high marks and restored its safety certifications. The gas division, Lloyd's said, had become a best-in-class operator.

Meanwhile, PG&E's electric system received no such assessment. The division seemed to be doing the right things, at least on paper. Spending on power lines big and small had risen under Earley's watch, even with so much attention on the gas system. Managers were constantly proposing projects to improve the reliability of the system. The lines that served the most customers often got the most attention.

Stavropoulos suggested to colleagues in the electric division that they invite Lloyd's to have a look. The response: not necessary.

Seven

NO FREE LUNCH

California's golden summers and gentle winters are as much a curse as they are a blessing. The wet-dry cycles of a Mediterranean climate are prone to extremes. Scarcely any rain falls during the warmest months, only to inundate the state in late-autumn torrents. Sometimes, rain refuses to fall at all.

In 2007, a dry winter gave way to a dry spring. By summer, more than half the state faced serious drought conditions. Reservoirs shrank, greenery turned brown, and trees fell ill for lack of nutrients. With drought comes fire, a fact as old as time. But that autumn, another variable compounded the risk. The Santa Anas, the dangerous winds that originate in Nevada and Utah and jump the Southern California mountains, blew with unusual strength. In September and October, dozens of fires ignited and spread quickly throughout the region as sparks from a range of sources—an arsonist, an overturned truck, a child playing with matches—settled in bone-dry brush. The deadliest and most destructive was the Witch Fire, the massive blaze sparked by one of San Diego Gas & Electric's power lines.

A wet spring offered some reprieve, but it was short-lived. The summer of 2008 became one of the driest on record, with rainfall 76 percent below average. That June, an earthshaking thunderstorm hurled more than six thousand lightning bolts from sky to forest. Trees erupted in flames. A satellite passing over California beamed images back to Earth. This time, Northern California was on fire, barely visible through a blanket of gray smoke billowing from the Sierras and the wooded chaparral west of the mountains. Enormous fires had also erupted near Monterey and Santa Barbara, forcing evacuations from thousands of coastal homes. Arnold Schwarzenegger, still governor at the time, drove throughout the state to survey the destruction. In early July, more than 1,700 fires were burning, stretching firefighters to the limit. Schwarzenegger ordered four hundred California National Guard troops to assist them.

"One never has resources for 1,700 fires. Who has the resources for that?" Schwarzenegger said, standing outside a command post in Santa Barbara as a nearby blaze exploded to cover more than 8,300 acres. "Something is happening, clearly. There's more need for resources than ever before. It's fire season all year round."

In late 2008, a few months after the fire siege, the California Public Utilities Commission opened a proceeding with a simple objective: push the state's utilities to address the threat of fire. Power lines had caused only a small fraction of the 2007 and 2008 fires. But the enormity of the Witch Fire had startled the agency into action by demonstrating the destruction that can result from windswept power lines throwing sparks during inevitable periods of drought. Staffers, in their opening remarks, wrote that the commission should consider revising its regulations to "protect the public from potential hazards, including fires, which may be caused from electric utility transmission or distribution lines."

Like most every other CPUC proceeding, this one kicked off with ideas and remarks from all of the state's utilities, as well as many other stakeholders. Dozens of proposals began circulating. There was little agreement on which of the agency's regulations needed to change. The debate played out mostly in thick filings posted in a lengthy docket that few members of the public would ever have reason to see.

As the docket lengthened, the drought worsened. Winter snow and rain had fallen at levels well below normal, and, come February 2009, California's reservoir levels were at historic lows. Schwarzenegger declared a state of emergency. The state was in its third year of drought. Lack of water, Schwarzenegger said, only increased the risk of devastating fires. The state's last major drought had occurred in 1991. Since then, California had added nine million new residents, many of whom had settled outside its major cities in the wooded foothills most vulnerable to burning. Any fire had the potential to be deadlier, and more destructive.

Meanwhile, the CPUC proceeding on fire risk had already gotten mired in bureaucratic mud. Commission staffers, frustrated by what they saw as quibbling among the utilities, felt that time was of the essence. The Santa Ana winds would inevitably return in the fall, with potentially deadly consequences. Certain parties, the staffers wrote, "would have the Commission act like Nero fiddling while

Rome burned," adding that there was "no dispute of the devastating nature of fires linked to power lines."

In February, about a month before the staffers expressed their concerns, strong winds had swept the southern Australian state of Victoria. Power lines fell, igniting a fire that barreled out of control to raze towns and farmland outside of Melbourne. It was one of hundreds of fires that erupted throughout the state that day, killing more than one hundred and seventy people. The fire outbreak was the worst in Australian history. CPUC staffers saw it as a harbinger of disaster for Southern California, which was even drier than southern Australia.

Historically, Southern California has been hotter and drier than Northern California, and therefore at higher risk of fire. But the drought, and the spate of fires that swept the state in 2008, foreshadowed swift and lasting change. Historical patterns wouldn't hold on a warming planet. As California smoldered, scientists reached consensus that climate change would stress water supplies around the world, increasing the risk of drought and wildfire. The American West would be among the first regions to feel an acute difference.

For California, a hotter, drier climate wouldn't just mean less rain—it would also mean less snow falling high in the Sierras to pile up in the winter and melt in the spring. Snowmelt accounts for more than half of California's water supply, flowing into the many rivers that cascade down the mountains to feed reservoirs. That water also hydrates forests. Without it, drought-parched trees would stand no chance.

In 2011, the California sky finally opened up, ending the three-year drought. Conditions appeared to stabilize. The CPUC proceeding, meanwhile, was still plodding along. Staffers were debating whether to require every California utility to prepare wildfire risk–mitigation plans. PG&E's attorneys resisted the idea that the company should be held to the same standard as its southern counterparts. In a thick filing that year, they argued that fires simply weren't as frequent in Northern California. The company, they said, already had a number of contingency plans that acknowledged fire risk, at least to some extent.

"PG&E does not agree that it is necessary to require a formal plan specific to fire prevention," the attorneys wrote.

Nothing in the filing mentioned climate change.

Climate change was top of mind for the CPUC's five commissioners. But they weren't deep in the weeds with their staffers, debating what the agency should do about the prospect of more drought- and wind-driven fires. Their focus was on making California a leader on climate policy, with an ambitious goal to make renewable energy account for 20 percent of the state's electricity supply by 2010. The commissioners, appointed by the governor, were tasked with holding the utilities accountable on the target and approving their contracts to purchase solar and wind power.

Michael Peevey, who had been president of the commission since his appointment in 2003, had made good on his promise to become the greenest president the commission ever had. Under his watch, PG&E had become a leader in renewable energy, having contracted for 8,200 megawatts by 2010. Southern California Edison and San Diego Gas & Electric had each secured large amounts, too. The utilities together spurred a massive regional build-out of wind and solar farms, the scale of which would help drive down future project costs.

In 2011, California's goals got even more ambitious when Jerry Brown, a famous figure within California politics, again took gubernatorial office. He had already been elected governor twice before, in 1974 and 1978, around the time of the widespread fuel shortages sparked by America's intervention in the Arab-Israeli war. He had been quick to establish California as an early leader on energy policy by imposing efficiency standards for appliances and buildings. It proved to be a more effective way to conserve than Nixon's daylight saving time extension.

Some thirty years later, Brown replaced Schwarzenegger, who had dramatically heightened California's focus on climate and the environment. Before leaving office, Schwarzenegger had advanced the idea of requiring renewable energy to supply 33 percent of the state's electricity sources by 2020. But drafting the legislation proved complicated, and Schwarzenegger hadn't seen a version he liked before he left office. Brown inherited the effort. In April 2011, three months into the job, he signed the bill. "Instead of taking oil from thousands of miles away, we're taking the sun," Brown said before the signing ceremony at a solar manufacturing facility in Milpitas.

The new goal meant that each of the state's utilities would have to procure even more wind and solar power, which at the time was substantially more expensive than other forms of generation. Already, renewable energy contracts cost them billions of dollars a year. Those costs fell to consumers. Rates for PG&E's residential customers rose an inflation-adjusted 9 percent between 2004 and 2010, to 15.8 cents per kilowatt-hour.

It soon became clear that the commission would approve even the priciest of renewable energy projects in the name of climate action. In November 2011, the commissioners were set to decide on whether to approve a twenty-five-year contract PG&E had negotiated to buy electricity from a 250-megawatt solar farm about 115 miles northeast of Los Angeles. A Spanish conglomerate called Abengoa SA was developing the project, known as Mojave Solar. It wasn't just any solar farm. It was a solar concentrator. Once complete, it would produce power with futuristic rows of mirrors covering two square miles of desert sand.

Solar concentrators are among the most expensive solar generators. Instead of solar panels, they use concentric arrays of mirrors that focus the sun's rays on a central point. The concentrated heat is used to power steam turbines. Some store the heat using molten salt and produce power even after the sun sets. They appear almost extraterrestrial in their flashiness.

Such projects had gotten a boost under President Barack Obama's administration, which had expanded a loan guarantee program run by the Department of Energy to include carbon-reduction technologies including solar concentrators. The office had agreed to a $1.2 billion loan guarantee to help finance Mojave, calling it a prime example of the sorts of projects needed to "compete with countries like China in the global clean energy race."

It wouldn't come cheap. CPUC staffers crunched some numbers and determined that Mojave's price per megawatt hour would be significantly higher than other sources of power. The staff ultimately recommended either striking the contract or renegotiating it. "For all the strengths underlying the Mojave Solar project, it has one significant weakness—the cost," the staff report read.

Peevey, who, as president, had the most influence of any commissioner, wasn't deterred. He liked the idea of the project, which would add a new kind of solar power to PG&E's portfolio and demonstrate the state's commitment to greening the grid. Plus, it had federal dollars behind it. Peevey became intent on its approval and secured the support of the other commissioners—except for one. Mike Florio, who had been appointed to the commission earlier that year, stood staunchly in the way. Florio, a friendly and plainspoken lawyer with big glasses and a full gray beard, had spent more than thirty years as an attorney for The Utility Reform Network, a consumer advocacy group that frequently intervened in CPUC proceedings when utilities asked for permission to raise rates. Florio sided with the commission's staff, arguing the contract would saddle customers with an undue burden.

The commission was scheduled to vote on the project on November 11. In the days before the meeting, Peevey pressured Florio to change his mind, sardonically calling him "Mr. No" right to his face. Brown's chief of staff, as well as a PG&E executive, each called Florio late at night in direct appeals for his support.

At the meeting, in remarks before the vote, Peevey emphasized that Abengoa had spent five years and $70 million to get the project ready for construction. If the commission were to reject it now, he said, it would send a "chilling message" to businesses and investors looking to bid new projects to support the state's clean energy targets, known as renewable portfolio standards.

"Future RPS bidders would be reluctant to invest the time and money needed to bring successful bids before the commission if they feel us to be so cavalier about rejecting otherwise viable projects," he said.

Florio wouldn't fold. From the dais, he marveled at the sheer cost of the solar concentrator. It would cost PG&E's customers $1.25 billion over twenty-five years, if not more. That worked out to at least $50 million a year. It wasn't that Florio opposed the state's renewable energy goals. His concern, he said, was that strapping utility customers with above-market costs would ultimately undermine or even defeat those goals by driving up rates. That would leave the utility less leeway to make other system investments, and give customers more reason to leave by joining one of the community-choice aggregation programs that Peter Darbee had fought so hard against. All that, Florio warned, could give way to a utility "death spiral."

"I cannot in good conscience vote for such an expensive—really expensive—renewable project," he said. "We have plenty of time to obtain less expensive, readily available renewable energy from other sources."

The commission approved the project four to one. The exact terms of that contract and others remain a secret, but a state report later revealed that PG&E agreed to pay an average of 19.4 cents per kilowatt-hour for power from four solar concentrators, including Mojave. That was the highest price paid by any California utility for any form of renewable energy.

By 2012, PG&E was spending more than $1.2 billion a year to meet its renewable targets, more than double what it spent in 2003. That spending was projected to top $2 billion by 2015, as a slate of new projects began producing power. Those costs would fall to customers in the form of higher rates.

State lawmakers saw reason to worry. In August 2012, Peevey appeared before the senate energy committee to address concerns that Californians faced the prospect of a "rate bomb" in the coming years. He explained that the contracts, though expensive, were making California a green energy leader, and costs for future projects would fall. (He was right—within a decade, wind and solar farms would become some of the least expensive forms of power generation, in part because of California's early investments.)

"There is no free lunch here," he told the senators. "I would hope that everybody by now accepts that climate change is real, and man contributes to it."

In 2012, hardly any rain fell in California. Drought again overtook the state with unusual speed and severity. Scientists said, unequivocally, that it bore signs of a changing climate. Governor Jerry Brown declared a state of emergency. He pointed to scientific projections that things would only get worse as the planet warmed.

The dry spell created unprecedented risk in California. Tens of millions of coniferous trees, already stressed after the drought just two years earlier, would soon die without water, creating an ideal environment for the invasive bark beetle. Bark beetles, smaller than grains of rice, burrow into tree trunks and feast on their nutrient-rich interiors. Healthy trees repel the invaders with a sticky resin that protects their bark. Thirsty trees produce less resin, making it easier for beetles to sap them of nutrients. The beetles mounted their assault as drought again set in. Northern California's lush green forests would soon become reddish-brown tinderboxes.

Meanwhile, after three and a half years of back-and-forth, the CPUC ordered Southern California utilities to prepare plans to address wildfire risk. San Diego Gas & Electric was ahead of the game. Since the Witch Fire in 2007, the company had been working on a number of efforts to reduce the risk of fire. It began taking careful stock of the trees near its power lines, tracking wind speeds with weather-monitoring stations and using new types of data to improve power line safety. The CPUC's decision required Southern California Edison to take similar steps. It did not, however, require the same of PG&E.

"Unlike Southern California, the need for electric utilities to develop fire-prevention plans in Northern California is not clear cut," agency staffers wrote. "To our knowledge, there has never been an instance in Northern California where strong winds have caused power lines to ignite large-scale wildfires."

The commission instead directed PG&E to "make a good faith effort" to determine whether there was a "credible possibility of extreme fire-weather events" in its sprawling service territory. Only then would the company have to draft a fire-prevention plan.

Northern California has its own version of the Santa Anas, called the Diablo winds. They're borne of the same circumstances: pressure building over Nevada and Utah, churning air up and over the California mountains. They sweep down Sierras in the north, just as they do the ranges near Los Angeles and San Diego. The hot, dry air pushes through the forests as it races toward the coast, cutting through San Francisco's iconic fog and dissipating over the Pacific.

Fire danger is highest when the Diablo winds blow, typically in autumn. They can lift sparks from any source—a campfire, a discarded cigarette, a burning pile of leaves—and drop them on the flammable forest floor. Northern Californians experienced their devastating potential in 1991, when the winds transformed a small grass fire into a deadly conflagration one weekend in October at the height of a severe drought. The fire spread through the hills just north of Oakland, destroying more than three thousand homes. Twenty-five people died.

In its 2012 decision, the CPUC required PG&E to identify power lines in the path of the Diablo winds and then calculate the probability of sustained gusts strong enough to seriously stress that equipment. PG&E took almost a year to do the study. It decided that the likelihood of such gusts was low enough to skip the full-blown wildfire plan. The conclusion had several fundamental flaws. It didn't directly account for branches falling or flying into power lines when the Diablo winds blew. And it assumed that the power lines were well maintained, capable of withstanding strong winds.

The last assumption was hugely consequential, and neither PG&E nor the CPUC was in a position to say it conclusively.

The CPUC's intense focus on climate policy came at the expense of one of its core responsibilities: holding the utilities accountable on safety. The San Bruno explosion had forced the agency to acknowledge its failure to do so. Its reckoning started with the National Transportation Safety Board report, which had slammed the agency for neglecting to detect problems with the way PG&E maintained its gas system.

Mike Florio was quick to point out the agency's shortcomings upon his joining the commission a few months after the explosion.

"We all have to admit that safety was something that was too much taken for granted," he said. "You can't do it without people—we need to beef up our staffing."

It was easier said than done. The CPUC brought in a consultant to interview staffers about why the agency's safety efforts had fallen so short. The consultant issued a report with a simple conclusion: The safety division received less money and staffing than the ones focused on renewable energy and setting rates for gas and electric service. "There has been little attention and limited resources directed toward reliability, and even fewer toward safety, by the Legislature and the Commissioners," the report stated. "Because safety is considered to be 'off the radar screen' of most Commissioners and legislators, it is considered to have little cache for CPUC staff and managers."

At the time, the CPUC's safety and enforcement division was understaffed with only about thirty employees. It mainly focused on auditing utility records, rarely going out in the field to examine electric or gas lines. The small staff simply wasn't large enough to run a meaningful inspection program. It had little choice but to trust that the utilities were properly conducting the work.

The CPUC then created a new seven-person team to assess risk and push the utilities to be more forthcoming when they discovered problems. But the safety division, which had long labored to fill vacancies, had a revolving-door problem. Agency jobs paid less than industry jobs. A number of its safety auditors moved into roles at PG&E and other utilities to oversee the functions they once regulated.

Peevey acknowledged that the agency had difficulty focusing on safety. The problem, he would later recall, was that it simply wasn't glamorous. It was difficult to get the legislature excited enough to expand the agency's safety budget, and more difficult still to staff that department effectively.

Even Peevey sometimes found it hard to engage when PG&E attempted to demonstrate its commitment to change. In April 2011, around the time Darbee resigned, the company announced that Jack Keenan and Ed Salas, who each held high offices within the utility, would also resign, as it attempted to clean house and reorganize after San Bruno. Geisha Williams, who had shown aplomb in addressing San Bruno residents after the explosion, would take charge of electric operations.

"Didn't realize Keenan and Salas were the problems," Peevey wrote to Brian Cherry, the regulatory relations head. "Who is Geisha?"

Cherry replied that she had joined PG&E about four years earlier with a background in electric transmission operations. She had been overseeing electricity delivery, he said, as well as some aspects of gas system maintenance and construction.

"Sounds like a classic reshuffling of the deck chairs," Peevey responded.

The CPUC, with its small safety staff and much larger policy team, was blind to an overwhelming problem snowballing within PG&E. The company, after changes to its business driven by deregulation, bankruptcy, and the transformation attempt, had cut back on certain maintenance and inspection work, lost critical records, and failed to effectively organize others. That made it increasingly difficult for the company to determine which of its power lines should receive the most attention. PG&E depended largely on reliability data to gauge the health of its system, and by those metrics, things had been improving. Outages were getting shorter and less frequent, in part because the company had spent time installing technology to automatically restore power to its lines after brief faults.

Reliability statistics reflect the number of customers affected by outages. For that reason, PG&E often prioritized urban and suburban lines serving millions of Bay Area residents. It paid far less attention to thousands of miles of lines that stretch deep into the Sierras, crisscrossing dry forests in the path of the Diablo winds. Many had been built decades earlier to support the hydroelectric system and meet demand growth after World War II.

Much of it was fast approaching the end of its useful life. In various filings over the years, PG&E had flagged to state regulators that its power lines and other equipment were aging, putting them at risk of breaking down more frequently. That was especially true of the transmission system, with hundreds of prewar towers and wires. A PG&E executive once said the company would have to process many projects, all at once, to prevent system failures—a problem he likened to a "pig in the python."

A few years earlier, PG&E had commissioned a consulting firm called Quanta to assess the age and condition of all of its transmission towers. It immediately discovered the same problem Accenture had encountered during the transformation effort years earlier: PG&E didn't have complete records of its system. The firm did its best with the data available, dividing up the lines by voltage class for assessment. PG&E's largest lines support 500 kilovolts—the equivalent of transmission superhighways. They get smaller from there, stepping down to 230 and 115 kilovolts. PG&E's 230-kilovolt lines are decades old. Quanta determined that about 60 percent of the towers supporting that network were built between 1920 and 1950. The 115-kilovolt lines, however, are the oldest of all, dating back to when Great Western and PG&E were still rivals. Quanta, in its assessment, was unable to determine the age of about 6,900 towers supporting them. It found that nearly 30 percent of the remaining towers—more than 3,500—were installed in the first two decades of the twentieth century.

The CPUC wasn't the only regulatory body responsible for monitoring PG&E's transmission system. Though the failure of deregulation in the early 2000s resulted in California's utilities retaining their roles as electricity retailers, the effort permanently muddied regulation of their transmission lines, the foundation of the state's wholesale power market. In the 1990s, the Federal Energy Regulatory Commission assumed oversight of transmission systems in several large regions to ensure the grid remained reliable as more companies became involved in producing and selling power. In doing so, it assumed responsibility for setting utility rates for transmission service. It mostly left maintenance issues to state regulators.

For an agency like the CPUC, with limited safety resources, transmission-line maintenance was challenging to monitor. California utilities had thousands of miles of high-voltage wires running throughout the state. Though the agency has hundreds of pages of rules for many aspects of power line operations, it's largely up to the utilities to determine how and when to inspect their equipment. The CPUC's rules for distribution lines delivering power to homes and businesses include some inspection requirements. The rules for transmission lines, meanwhile, are just three sentences long. They simply say that each utility must come up with its own procedures and follow them.

There is another body involved in transmission regulation: the North American Electric Reliability Corporation, the FERC-controlled nonprofit that took on greater importance after the Northeast blackout of 2003. The blackout, which swept from New York to Canada when trees touched a midwestern utility's transmission lines, raised concerns that other utilities weren't doing enough to keep their lines in shape. In 2010, NERC, as the body is known, directed utilities across the country to take stock of their lines and make sure the wires were the proper distance from the ground and trees and from the towers holding them aloft. From NERC's standpoint, this was primarily a reliability issue. But it was also a safety issue. High-voltage wires too close to vegetation might not only trip off, but also cause fires. PG&E workers fanned out to see what lines needed work.

The Caribou-Palermo, the century-old transmission line that would later spark the Camp Fire, was among them. In 2013, PG&E told federal regulators it planned to replace many of the towers, wires, and hardware pieces on the line. It spelled out the scope of the work in emails to federal forest managers, telling them it needed to replace forty-nine steel towers "due to age," and hardware and conductors on fifty-seven towers "due to age and integrity." It wouldn't have addressed the tower where the Camp Fire started. But it would have made other parts of the line safer.

For reasons unclear, PG&E delayed the project. It would be delayed again, and again, and ultimately never completed. It was one of many transmission line projects that got pushed back, year after year, as the company failed to recognize just how risky its territory was becoming. Many of the projects were in the Sierras.

Meanwhile, the CPUC launched an effort to map high-threat fire areas throughout the state. Staffers began studying how drought and climate change were affecting the health of the forests surrounding the Central Valley. They began marking the riskiest areas in yellow and red. It would take years for them to determine that almost all of the forests, stretching from Southern to Northern California, were yellow, with ominous spots of red throughout Napa, Sonoma, and the Sierra foothills.

In 2014, public concern about the CPUC's ability to regulate a company like PG&E crescendoed when a cache of incriminating emails came to light. In the process of suing PG&E, lawyers for the City of San Bruno requested thousands of pages of emails and correspondence from the company's officers. Huge batches of records held evidence that CPUC commissioners had engaged in back-channel communications with PG&E executives, which was banned under the agency's rules. Their emails, which were made public, raised serious questions about whether the commissioners could effectively police the company.

Almost all of the commissioners were implicated to some degree. One of the most infamous exchanges occurred four months before San Bruno, when Cherry invited Peevey to his house for dinner. "No matter the menu," he wrote, "we have some great bottles of Pinot to drink."

To outsiders, especially consumer groups like the Utility Reform Network, the chatty emails and boozy dinners seemed to have all the hallmarks of an old boys' club. Critics of the agency accused it of helping PG&E get exactly what it wanted. Some emails suggested that PG&E had asked regulators to look the other way as it sorted through some of its safety problems after San Bruno. One of the most damaging exchanges showed PG&E criticizing the judge assigned to oversee its request to increase gas system spending at its customers' expense. Cherry, the regulatory relations head, panicked when he learned of the assignment.

"This is a major problem for us," he wrote to Peevey's aide. "I'm not sure we could get someone worse. This is a very important case that is now in jeopardy."

He ramped up pressure on the agency to change the assignment, which it ultimately did.

PG&E fired Cherry and two other executives when the communications came to light. The scandal also ended Peevey's tenure. He didn't seek reappointment when his second term as president ended in 2014. After an investigation, the CPUC eventually fined PG&E $97.5 million for improper communications with its own officials.

Peevey never saw it that way. In his view, such friendliness was necessary to get things done. Part of the role of the regulator is to negotiate and occasionally compromise, and Peevey didn't see much value in being antagonistic in that process. Hobnobbing was part of the way he struck deals and developed relationships with the people he had to work with.

Still, the damage was done. The commission now had to contend with a reputation for being dangerously close to the utilities it regulated. In 2015, Governor Brown appointed Michael Picker, a longtime political consultant who had joined the commission the prior year, to succeed Peevey. A trim man with flyaway white hair and a thick mustache, Picker had been focused on California's energy goals for years, having advised both Brown and Schwarzenegger on renewable projects and policy. He wasn't a natural dealmaker. The idea of settlements, he once said, made him queasy, for fear someone had gotten short shrift in negotiations. He immediately drew a hard line, saying that PG&E had been banned from communicating with anyone within the agency except through proper public channels. There would be no more boozy dinners or chatty emails.

"If there's a prohibition on having a conversation, I just don't have it," Picker said during his first remarks as president.

He called the incriminating emails "troubling and very painful to read" and said the CPUC was cooperating with federal and state investigators. Then he offered a more somber assessment of the agency's competence. He listed what he saw as some of the biggest problems: no consistent practices for safety enforcement and investigations. The National Transportation Safety Board, he reminded the commissioners, had been almost as hard on the CPUC as it had been on PG&E in spelling out the many lapses that led to the San Bruno explosion.

"What keeps me awake in the early hours is not emails. It's the slow erosion of our safety programs," Picker said. "That is the thing that truly shakes me."

The stakes were mounting. Around the time of Picker's appointment, forest officials convened a drought task force to deal with dying trees. It was both a state and federal issue, given the number of national forests throughout the Sierras. Forest managers began considering how to thin them by removing young trees and flammable brush to reduce the likelihood of enormous, uncontrollable wildfires. CPUC staffers, meanwhile, were still debating fire risk. That year, the agency required the utilities to begin reporting every fire started by their equipment. PG&E wasn't keen to do it, but it began keeping a spreadsheet. In the final seven months of 2014, it tracked 254 fires. Most were small and quickly extinguished. But the number would only continue to rise.

PART

II

Eight

THE TRIAL

James Haggarty would never forget the night of September 9, 2010. It was his thirty-third birthday, and he had dinner plans with his family to celebrate. They had just walked into a restaurant in Burlingame, south of San Bruno, when they heard the sound of a siren. Another followed, then another, and another. Haggarty, an officer with the San Bruno Police Department, knew that many sirens could only mean disaster. He walked outside and looked north to see a fireball in the distance, its flames licking the sky. His bosses called him into work. The fire department needed officers to secure the area. When they relayed the location, Haggarty knew immediately where it was. He had been to the exact spot countless times. The pipeline spewing gas and flame had exploded just two blocks from his childhood home.

The US Attorney's Office for the Northern District of California began investigating PG&E shortly after the explosion, and the San

Bruno Police Department assigned one of its detectives to the case. When that detective took on a new role in early 2012, it needed someone to replace him. Haggarty, who had been promoted to detective just a few months earlier, had been intrigued by the case. For him, it was personal. And he loved the idea of a challenge—white-collar crime was a far cry from the homicides and street gangs he usually investigated. He got the assignment.

Haggarty, tall and broad shouldered with dark wavy hair, took the case knowing next to nothing about pipelines. The first week, he recalled, was like drinking from a fire hydrant. To hand off the assignment, his colleague produced binder after binder filled with dense material. Haggarty, at an early meeting with regulatory attorneys, could barely make sense of the conversation. He left determined to get up to speed.

He began in the library, where he taught himself all he could about gas transmission. A subject matter expert then gave him what amounted to a college course on the topic. Within months, Haggarty had become one of the prosecutors' greatest assets. He took the case head-on, probing every possible lead until he had mapped out decades' worth of the company's maintenance and engineering decisions. Part of it involved interviewing dozens of current employees, former employees, regulators, and consultants.

One interview was especially revealing. On September 5, 2012, Haggarty called to interview a consultant who, at the direction of the California Public Utilities Commission, had audited PG&E's gas system spending between 1997, around the time of deregulation, and 2010, when the explosion occurred. During that period, the consultant found, PG&E had spent $39 million less than it told regulators it would spend on pipeline operations and maintenance. At the same time, it exceeded its authorized rate of return by several hundred million dollars by investing heavily in capital projects. The maintenance spending cuts were particularly acute between 2008 and 2010. The consultant found that engineers had deferred numerous safety projects to meet budget targets set by management.

"Did it seem odd to you that that company was making more money, while reducing spending on safety?" Haggarty asked.

"It didn't make sense to me," the consultant replied. "Why the extreme budget pressures within these years? That's the good question."

A couple weeks later, Haggarty sat down to interview a manager within the gas transmission division who had the answer. The manager had worked within the division for nearly thirty years and had overseen its budget for a decade. Come 2007, his team was under mounting stress. Business Transformation was faltering, incurring more costs than it offset. That same year, employees appeared at the annual shareholders meeting to flag concerns about leaks of the sort that resulted in the Rancho Cordova explosion. The company needed to ramp up spending within gas distribution—the system of smaller pipes delivering gas to homes and businesses. In order to fund higher expenses without compromising profits, upper management decided that gas transmission would have to further cut costs. The division had requested $120 million in 2008. Regulators approved less than that, a common occurrence when the company requested permission to recover expenses through rates. They signed off on roughly $95 million. PG&E then decided the division would receive even less money than that—only about $90 million.

In late 2007, the manager overseeing the gas transmission budget fired off an email to his team. The subject: "Urgent: Minimum funding levels for IM 2008-2010." IM referred to integrity management, the program for inspections and maintenance. He warned that the division would "be receiving very low funding for expense work," and that cuts would be needed.

"The optimal level of funding will be well below what is necessary to properly run the business," he wrote.

The prosecutors empaneled a grand jury to decide whether PG&E should face criminal charges. The company responded to subpoena after subpoena with more than ten million pages of documents. It took months for Haggarty to sort through all of it. He and his team worked nights and weekends to build a case against the company, picking out thousands of emails, spreadsheets, and memos that circulated among employees in the years prior to the San Bruno explosion. The grand jury, after reviewing the mountain of evidence, reached a conclusion: PG&E had acted with conscious disregard for safety in maintaining its gas transmission system.

In April 2014, prosecutors charged PG&E with twelve counts of violating federal pipeline safety laws. All were felonies. The indictment went well beyond the problems with the pipeline that exploded. PG&E, it alleged, had a history of treating the law with knowing defiance. The prosecution didn't buy the idea of simple negligence. It set out to prove the company's criminal intent.

The prosecution's argument centered on the 1968 Natural Gas Pipeline Safety Act, which required pipeline operators to maintain thorough records on their systems. The grand jury found that PG&E lacked records related to almost every aspect of its gas operations: procurement, manufacturing, construction, pressure testing, inspections, and repairs. Those it did have were riddled with inaccuracies. Employees, regulators, auditors, and consultants had all sounded warnings over the years, but the gaps persisted. In the 1990s, the company had attempted to consolidate its records into a central database. Employees used it to plan maintenance and inspections. Unsound records made for unsound decisions—many documents were still stored as hard copies in scattered repositories. (The company would later consolidate them all at Cow Palace in the frantic records search after the explosion.)

In the early 2000s, Congress amended the law to require pipeline operators to survey their systems for potential safety threats and assess them with certain testing and inspection methods. PG&E, with its faulty database, couldn't identify all of the risks. Employees knew it. Some of them raised concerns. But the company, facing budget pressures, failed to gather more data and then went on to cut corners on inspections and testing, the indictment alleged.

Each of the charges began the same way: PG&E, through the actions of its employees, "knowingly and willfully" violated the law.

On April 21, 2014, PG&E pleaded not guilty.

PG&E's investors were apprehensive. It wasn't immediately clear what a conviction would mean. The charges were against the company itself, not individual employees, so no one would go to prison. Some sort of fine seemed inevitable, as did another reputational hit. Tony Earley, the company's CEO since 2011, tried to allay concerns during an earnings call a couple of weeks after the company entered its plea. Regarding the penalty, Earley said, the "dollar figures are not big issues here." If convicted, the company wouldn't have to pay more than about $500,000 per count, for a total of $6 million—less than 1 percent of the company's profits that year. And Earley didn't expect that to happen.

"Fundamentally, it's our belief that the criminal charges against the company just are not merited," he said.

Meanwhile, Haggarty had been contemplating another charge. Months earlier, while sorting through the mess of documents, he came across records and correspondence showing that PG&E had obstructed the National Transportation Safety Board's investigation of the explosion by lying about the way it cut corners. He determined that the company had knowingly established a dangerous policy for inspecting pipelines after pressure increases, and he believed that he had the evidence to prove it. The grand jury, after reconvening for one day, agreed.

In July, the prosecutors filed new charges against the company. They alleged that the company had not only violated the Pipeline Safety Act but also obstructed the NTSB's investigation, casting doubt on whether it had acted in good faith in its cooperation. And the superseding indictment was broader in scope, counting additional problems on different pipelines as separate charges. Suddenly, the company faced one count of obstruction and twenty-seven counts of violating pipeline laws.

Earley appeared before the *San Francisco Chronicle*'s editorial board shortly after the new charges were filed to face questions about the company's culpability. It was clear that Earley, a lawyer himself, understood that the fate of the company depended on whether the prosecutors could prove the willful misconduct of its employees.

"Did people make bad judgments? Yeah, they made bad judgments," he told the board. "But did they say, 'I know this is what the law requires—I'm just not going to do it that way?' We haven't seen any evidence of that, or at least I haven't."

Leading the prosecution were Hallie Hoffman, Hartley West, and Jeff Schenk, assistant US attorneys for the Northern District of California. They were all within a few years of each other, in their late thirties and early forties, and for all their different case experience, they had one thing in common: this one was the most consequential they had tackled in their careers.

Prosecuting a corporation is a special challenge within the American legal system. It assumes that a corporation is a person that can act with consciousness and intent. That assumption is hardly intuitive. Corporations act through their executives and employees—at what point does the action of an individual become the action of the entire organization? Lawmakers have struggled with the question for centuries. In eighteenth-century England, amid the rise of the East India Company and other aggressive multinationals, Lord Chancellor Edward Thurlow felt hamstrung in considering whether a corporation could be convicted of a crime. "Corporations have neither bodies to be punished, nor souls to be condemned; they therefore do as they like," he said.

In the United States, where corporations have wielded power since the charter of the British colonies, the concept of corporate personhood is rooted in a United States Supreme Court case heard three decades after the ratification of the Constitution. The 1809 decision, involving Alexander Hamilton's Bank of the United States, determined that corporations have the constitutional right to sue in federal court. The precedent set the stage first for the recognition of corporations as people under the Fourteenth Amendment, and then for the expansion of their civil rights to include many of those afforded to individual citizens. It remains an evolving subject, with two recent Supreme Court cases—*Citizens United v. Federal Election Commission* and *Burwell v. Hobby Lobby Stores*—thrusting the controversial idea of corporate personhood into the public consciousness.

The idea of corporate liability, meanwhile, is almost as old as PG& E, stemming from a 1909 Supreme Court case involving a New York railroad company. The court's decision set a significant precedent: A corporation could be criminally prosecuted for the actions of a single employee. In the court's view, corporations had unduly benefited from "the old and exploded doctrine that a corporation cannot commit a crime." Corporations had grown in number and influence in the decades since Thurlow lamented their lack of body and soul, with monopolies and trusts wielding an unprecedented amount of economic power. If corporations couldn't be held "responsible for and charged with the knowledge and purposes of their agents . . . many offenses might go unpunished," the court said.

That raised a thorny question. How should a corporate criminal be penalized? For decades, there were few choices besides financial penalties, with fines rarely exceeding a few hundred thousand dollars. In the 1990s, a federal sentencing commission issued sweeping new guidelines that enabled courts to collect larger fines as well as monitor convicted companies by putting them on probation and requiring them to develop compliance programs. The goal was not just deterrence but also rehabilitation.

In filing the new charges against PG&E, prosecutors sought a far larger penalty than the one outlined in the original indictment. The statutory maximum fine for violating the Pipeline Safety Act was $500,000. If convicted on all counts, the company wouldn't have to pay more than $13.5 million, still just a fraction of its annual profits. So the prosecution sought what's known as an "alternative fine." Such fines are calculated based on estimates of how much money a company gains from its misdeeds, or how much those misdeeds cost the public. The prosecution offered calculations for both. It suggested PG&E should have to pay as much as $1.13 billion.

Hoffman, Schenk, and West had their work cut out for them. A conviction would require proving, beyond a reasonable doubt, that PG&E's employees had both knowledge of the law and the intent to break it. Extracting the larger fine required the same standard. The team would have to show that its complex calculations of the gains and losses resulting from the crime were indisputable.

The day after prosecutors filed the new charges, Earley again spoke to investors on a quarterly earnings call. The indictment was among the first items he addressed. He reiterated his belief that the charges were unwarranted. Analysts asked about the possibility of a settlement. A company attorney said talks weren't underway. One pressed Earley on why the company wasn't trying to negotiate one and move on, given the prospect of a drawn-out case that could spook investors for years.

"There's no evidence that we've seen that somebody willfully and knowingly violated the Pipeline Safety Act," Earley repeated. "And if we admitted that we have willfully and knowingly violated that Pipeline Safety Act, that could have had consequences in the ongoing proceedings that we have."

At the time, PG&E was dealing with a host of other legal issues. In late 2013, it had agreed to settle the majority of the civil lawsuits seeking compensation for injuries and property damage resulting from San Bruno. The company had agreed to pay about $565 million; any admission of wrongdoing could have affected its ability to recover some of the cost through its insurance policies.

On top of that, the California Public Utilities Commission was debating how to punish PG&E for its negligence leading up to San Bruno. Its investigation predated the appointment of Michael Picker, who had inherited the effort from Michael Peevey. Within the CPUC, the commissioners themselves aren't involved in investigations. That's left to staffers, who confer with administrative law judges tasked with hearing testimony from all parties interested in the case. A month after the earnings call, the judges involved in the San Bruno case issued a recommendation. They proposed fining PG&E $1.4 billion for 3,798 violations of state and federal laws related to pipeline safety. For PG&E, it was a frighteningly large figure. The company issued a public plea asking the CPUC to reduce the penalty to something "reasonable and proportionate." Picker was unsympathetic. The following April, he proposed not a reduction but an increase, raising the fine to a record $1.6 billion. The commission voted unanimously to impose it. The previous record had been $38 million, levied against PG&E for the Rancho Cordova explosion.

In corporate America, few large companies are federally prosecuted. Fewer still go to trial. The overwhelming majority settle, often with what's known as a deferred prosecution agreement, in which the Department of Justice or the Securities and Exchange Commission holds off on pressing charges so long as the corporation admits some level of wrongdoing, commits to reform, and perhaps pays a fine.

There have been notable exceptions. In the early 2000s, during Enron's meltdown, another company was pulled into the fray. Federal prosecutors charged Arthur Andersen, Enron's accounting firm, with one count of obstruction of justice. In the weeks after it received notice that the SEC was investigating Enron, the firm shredded thousands of documents and trashed emails that could have been used as evidence of Enron's financial deceit. The question was whether employees were simply following the firm's vague document retention policy, which required periodic purges, or attempting to hide its culpability in the Enron mess. Prosecutors alleged the latter. In their words, "employees were told to work overtime, if necessary, to finish the job of destroying documents. The shredder at the Andersen office in the Enron building ran virtually constantly."

The indictment was filed in March 2002. Arthur Andersen employees were furious, and frightened. A conviction would essentially be a death sentence, as the SEC would likely bar the firm from auditing publicly traded companies. On a warm overcast day, employees gathered outside of the federal courthouse in downtown Houston to protest the charges. Their black shirts were stamped with large red letters: I AM ARTHUR ANDERSEN.

The defense couldn't have asked for a better image to show the public the enormous challenge, and consequence, of convicting a corporation (or in this case, a partnership). Though the law recognized it as such, Arthur Andersen wasn't really a person in its own right. It was its employees. If just one of them obstructed justice, the firm could be convicted. But who did it, and how? The success of the prosecution's case hinged on finding at least one "corrupt persuader":

an employee who acted knowingly and intentionally to keep documents out of the government's hands by persuading another to destroy them.

Arthur Andersen's lead defense counsel, a caricature of a lawyer who dressed in bright, colorful suits and exasperated prosecutors with courtroom antics, used his opening remarks to remind the jury that in order to convict the company, a single employee had to have committed the crime. "You know the little routine Where's Waldo?" he asked. "Where is Waldo? Who are they?" The lawyer kept punctuating the point as Arthur Andersen employees took the stand as witnesses. "Are you Waldo?" he would ask. He sounded a warning to the jury: "When it's all over, you still won't know where Waldo is, and you still will not have found a corrupt persuader; you will just have a destroyed company."

After ten days of deliberation, the jurors agreed that they had indeed found Waldo and voted to convict Arthur Andersen. On appeal, the case went all the way to the Supreme Court, which overturned the conviction on the grounds that the prosecution hadn't properly educated the jury on the meaning of the word *corrupt*. But the damage was done. The firm had severed ties with its clients, closed its offices, and auctioned off the furniture. It was as if the court had reversed the sentence of a prisoner on death row after his capital punishment. Arthur Andersen was innocent in the eyes of the law, but it was too late.

The so-called corporate death sentence is exceedingly rare. Arthur Andersen was beholden to specific regulatory requirements because of its influence in the world of trading and finance. Even as convicted felons, most other companies can continue making widgets or selling services. That's especially true of a utility, whose services are critical. A utility can't shut off its power lines and let its

employees go; that would cause societal collapse. A utility's ownership and structure can change, but its primary function inevitably
remains the same. Convicted or not, PG&E faced no risk of death.

Still, the decision to go to trial wasn't unanimous among PG&E
executives. Many, including Earley, recognized how damaging a conviction could be, even if the company was sure to survive intact.
Whether the company should attempt to reach some sort of settlement had been the subject of intense debate. The company's lawyers
walked the executives through the options more than once. At the
federal level, settling a criminal case often requires some admission
of wrongdoing, and even a selective admission had the potential to
haunt the company. And then there was the question that hung over
the Arthur Andersen trial. Who was the bad actor? The executives
had sat down with a number of pipeline engineers to learn what had
happened. When the inevitable question came up—did you intend to
violate the Pipeline Safety Act?—the answer was always no. Waldo
hadn't made himself known.

The case was assigned to Judge Thelton Henderson, a legendary
figure who had forged his career fighting inequality and injustice. As
corporations were vying for expansive rights under the Fourteenth
Amendment, he took up the fight to secure those same rights for
African Americans, the people the amendment had been enshrined
to protect. Henderson had been one of two Black students in Berkeley
Law School's class of 1962, graduating to become the first Black
attorney in the Civil Rights Division within the Department of Justice. He embedded in the South, where he met the most prominent
activists of the day. One night, one of them had car trouble on his

way to the marches in Selma. He asked to borrow Henderson's government loaner, and Henderson agreed. The activist was Martin Luther King Jr.

The incident cost Henderson his job as critics complained that the government was too close to the activists. So he returned to California and worked in private practice and educational roles until 1980, when President Jimmy Carter appointed him to the federal bench. His most prominent cases involved the state prison system and the Oakland Police Department. Officers in both organizations had abused their power, treating inmates and suspects with neglect or brutality. These were agencies, not corporations, but they had similar characteristics, with hierarchal power structures and a wide diffusion of responsibility among employees. Henderson sentenced the organizations with the same means contemplated under the new guidelines for corporations, with requirements for reform and compliance and federal monitors to oversee the progress. Henderson was eighty years old when prosecutors charged PG&E. His mind was sharp, but an autoimmune disease had sapped his strength. The company's case would be among the last he heard.

PG&E's defense attorneys, meanwhile, were at the height of their careers. Leading the defense was Steven Bauer of Latham & Watkins LLP, a firm with decades of experience defending corporations. Bauer had just become head of the firm's white-collar defense practice. He had defended people and institutions accused of money laundering, insider trading, securities fraud, and tax evasion. He was also deeply familiar with PG&E, having worked with the company on various issues since the 1990s. Aggressive and unyielding, he was exactly the sort of attorney a company would want on its side.

Hoffman, West, and Schenk, meanwhile, had been holed up for months preparing for trial. They had lined their war room with floor-to-ceiling bookcases stacked with the millions of documents unearthed during discovery. They pulled all-nighters combing through the material, deciding what should be presented as exhibits. They needed to distill the evidence to its essence—confusion among the jurors could cost them the case. Early on, they used poster-sized paper to print the photos and life stories of the eight people killed in the explosion. They hung on the wall as a reminder of why the work mattered.

As with any indictment, the government's case rested on a certain set of arguments underlying the charges themselves. Bauer was determined to poke holes in each of them. In the summer of 2015, the Latham & Watkins team filed motions to dismiss every charge against the company. It offered reasons why each one should be thrown out on a host of technicalities.

Its first argument had to do with the statute of limitations. The prosecution had alleged that PG&E's record-keeping crimes began in the 1970s, not long after the pipelines were first built. Most federal crimes must be brought within five years. If any crime had been committed, the defense said, it was too late to do anything about it.

"While the government may wish to present evidence of record-keeping that occurred when rotary phones and 8-track tapes were still the norm, basic principles of law and fairness bar it from doing so," the defense wrote. "The alleged crimes are simply too far in the past and too remote to support a modern-day prosecution."

Next, it sought to dismiss every pipeline-related charge by arguing that the CPUC, not the federal government, had jurisdiction over PG&E, because the company's pipelines didn't cross state lines. Even if the company were to face federal charges, the defense argued, the prosecution shouldn't be allowed to count the same problems on different pipelines as separate crimes. "The government might just as easily have charged thousands of counts, treating every foot of pipe-line, every PG&E employee, or every page of the report as a sepa-rate crime," the defense wrote, referring to the NTSB's thick set of findings.

Next, the defense attacked the obstruction charge as baseless. Even if it had merit, the defense argued, it should be thrown out. Statute defined obstruction as the intentional impediment of a "pro-ceeding" before an agency convened to "administer" the law. The NTSB investigation was no such proceeding, the defense contended, given that the agency did not administer the law. By that logic, whether PG&E obstructed its investigation was legally irrelevant.

Finally, the defense went after the heart of the matter: the com-pany's "knowing and willful" mens rea, or state of mind. The prose-cution believed that employees' collective knowledge about their failure to follow the law could be used to prove the company's intent to break it. Precedent wasn't clear, and the defense seized on the ambiguity. If individual employees knew about various aspects of the pipeline problems, should that render the company criminally liable for their collective actions, or simply negligent for failing to provide them with the means to combine their knowledge to ad-dress the issues?

"At least one human being must intend to break the law in or-der to bind a company," the defense wrote. "The government cannot aggregate different bits of knowledge held by distinct individuals and argue that the organization as a whole acted with specific intent."

As the defense filed the last of its motions, the mercury began climbing in the Sierra foothills east of Jackson, a small town outside of Sacramento. It was 102 degrees on September 9, 2015, without a cloud in sight. Hardly any rain had fallen since Governor Jerry Brown declared the drought a statewide emergency more than a year earlier. The scrubby brush and grasses covering the hills were so dry they crackled. That warm afternoon, the spindly top of a gray pine tree snapped, falling onto a small PG&E power line feeding nearby residences. The branches ignited, dropping embers into the dry grass below. The fire spread to cover nearly seventy-one thousand acres, killing two people and destroying 921 structures. It took three weeks for fire crews to contain the blaze, which became known as the Butte Fire.

The reasons for the fire were a bit unusual. It wasn't the Diablo winds that had fanned the flames; there was scarcely a light breeze on the day the tree top cracked. The problem was the tree itself, which had grown within a stand of other trees that had protected it from the elements. Because of that, it was weak. It hadn't developed the trunk strength needed for stability. PG&E contractors had come by a few months earlier and removed the surrounding trees, leaving the weak gray pine exposed. It was only a matter of time before it gave out.

The fire's size and speed foreshadowed the grim challenges PG&E would face as its service territory grew drier and more flammable. Its constant effort to trim or cut down trees, known as vegetation management, was becoming a gargantuan task with critical consequences. Millions of trees were growing and swaying along thousands of miles of power lines. It would take only one limb touching one conductor to spark a wildfire.

The Butte Fire was hardly the only one in PG&E territory that year. The company's equipment sparked more than four hundred fires in 2015. Most of them were small and extinguished quickly, but the company nonetheless had to report all of them to the CPUC under the new requirements implemented a year earlier. The CPUC, meanwhile, paid little notice to the numbers, which were up sharply from the prior year. It was still trying to determine the areas where fire risk was highest. Three years into the drafting effort, the fire maps still weren't done.

A month after the Butte Fire, as Henderson considered the defense's motions to dismiss the charges, Earley put on a blue striped dress shirt and a dark suit jacket loose in the shoulders. He looked tired, his thick hair graying at the temples. He had agreed to tape an episode of *Climate One*, a public forum and podcast series held at the Commonwealth Club. The host, a former journalist named Greg Dalton, had launched the show a few years earlier as a platform to discuss environmental and policy matters with business executives, scientists, and other thought leaders.

There was reason to invite Earley on the show. Under his leadership, PG&E had continued to make progress on meeting its renewable energy targets even as it plowed money into pipeline repairs. By 2015, renewables were serving 30 percent of the company's electricity needs, putting it squarely on track to meet the state mandate of 33 percent by 2020. But it was starting to look as though Mike Florio, the CPUC commissioner, had been right to worry about expensive contracts. PG&E was paying an average of twelve cents a kilowatt-hour for renewable electricity. Four years earlier, it paid less than eight cents. Those costs were passed on. Adjusted for inflation, electric rates for residential customers were 14 percent higher than they had been a decade earlier. And they were beginning to feel it.

Earley and Dalton sat together onstage in wooden chairs. The cost of renewables came up, but only in passing. Dalton mainly asked about how the utility business model was changing as more homes added rooftop solar panels. Then he launched into a lightning round of questions posed as statements. Earley had two choices: Yes or no.

"Julia Roberts did a fabulous job playing Erin Brockovich in the movie about contaminated water and the California community, yes or no?" Dalton asked.

Earley smiled. "Yes," he said.

"The California Public Utilities Commission under former president Mike Peevey got a little too cozy with utilities it regulated," Dalton said.

Earley laughed nervously. "Boy, can I take the fifth on that one?" he said. "There were issues on both sides that really needed to be fixed."

"You are still working to repair the damage to the company's reputation caused by Peter Darbee's tenure as CEO," Dalton said.

Earley, face straight, didn't skip a beat. "Yes," he replied. That work was about to get harder.

A few days before Christmas, Henderson issued his rulings on the defense's motions. He denied most of them. The defense notched a single win. The judge agreed that the same problems on different pipelines shouldn't be counted separately in the indictment. The company would face only twelve charges, with a potential alternative fine of $562 million. But it was nonetheless headed to trial.

Jury selection began in June 2016. Dozens of people, mostly from the Bay Area, were summoned to the federal courthouse in San Fran-

cisco, a boxy gray building near the edge of the Tenderloin neighborhood. Those hoping for a speedy trial were soon dismayed. Henderson estimated it would last for six to eight weeks. The jury would
hear arguments Tuesday through Friday, 9:00 a.m. to 1:30 p.m.,
sometimes longer. Twelve people lined up to make their case for
excusal—job obligations, childcare, health problems. They were dismissed.

Those remaining were in for an intense screening. Unlike a criminal trial in which the jurors don't know the defendant, most everyone knew PG&E. Many of them had been longtime customers. And
almost all of them had heard at least something about San Bruno.
PG&E had already been tried by the media, with the nightly news
and local papers still churning out headlines about the explosion. It
would have been almost impossible to miss.

Another cloud hung over the defense. America was awash in anti-
corporate sentiment. Anger at big banks and businesses had crested
after the recession and plateaued during the plodding recovery. *Citizens United* and *Hobby Lobby*, the Supreme Court cases that took expansive views on corporate personhood, had been decided in the middle
of it all, fanning the flames of mistrust. Overturning *Citizens United*
became part of the rallying cry of the Occupy Wall Street movement,
the nationwide protest against inequality and corporate influence that
began in 2011 when several hundred activists descended on New
York's financial district.

The defense team told Henderson early on that the potential for
prejudice among the jurors was unusually high for a criminal case. It
pushed for a questionnaire to help screen for bias. All of the prospective jurors filled out sixteen pages before arriving in court. The questions cut right to the chase.

When you hear or see "PG&E," what is the first thing that comes to your mind?

Have you ever paid bills to PG&E? If YES, do you believe the bills were too high? Too low? About right?

Do you have negative opinions about PG&E?

The responses invigorated Bauer's interrogation process as jury selection began in court. One of the first to be questioned was an older woman who lived in a Bay Area condo. She had grown mistrustful of corporate monopolies after thirty-five years of working among the rank and file at AT&T. PG&E, in her mind, seemed like a similar beast. As with her phone service, she had little choice but to write PG&E monthly checks. The price she paid, driven higher by the company's renewable energy contracts, was a constant source of anxiety.

"I don't understand why my rates are still so high when I don't use my electricity, barely," she told the court. "And I call up, I go, 'Why is my bill now $65 when it used to be $40? You know, I'm on a budget here and I'm trying to figure this out.' And nobody has a reason."

She was dismissed.

The next prospective juror was a theater director from San Francisco who had followed the news coverage of San Bruno. He had heard a lot of talk about "deferred maintenance" and PG&E's negligence in allowing its pipeline system to deteriorate. He would have had difficulty putting that idea aside.

"I don't think that goes away just because I say I'm not going to have an opinion," he told the court. "If I was PG&E, I'd be worried about me as a juror."

He, too, was dismissed.

And so it went for nearly three full days. Bauer teased out all the anti-PG&E sentiment, asking the prospective jurors to give it to him straight. He thanked them for their honesty. Then he shifted his line of questioning. The sixteen-pager had included a series of statements framed to determine how the jury pool saw corporate America as a whole.

Large corporations are mainly interested in making big prof-its at the expense of others.

Most people with power try to take advantage of people like me.

Corporate executives will lie to win a lawsuit.

More than a dozen of the respondents agreed with those statements, at least in part. Bauer was ready to weed them out. "I saw a bunch of answers that said, I think, that corporations are taking advantage of me. The executives might lie for money," he said on the second day of the selection process. "I have to ask you about it because there is no secret about it, I'm representing a corporation here."

Someone piped up, a man who worked in computer-assisted design for Home Depot.

"Due to *Citizens United* and other things I have some very negative opinions of corporations," he said. "Anyone who knows me knows that."

Another man, a self-employed massage therapist, piled on.

"I do feel that some of the pay disparities between the top CEOs and the working people is pretty obscene," he said. "And there are some issues of accountability when corporations do things wrong."

A woman who owned a small apartment building in Oakland agreed.

"That's pretty much a trend for everyone nowadays, given, you know, wage disparity, *The Wolf of Wall Street*, the crash," she said. "It's kind of hard sometimes to see CEOs and stuff, you know, and take them seriously."

The pool shrank with each excusal. Eighteen were left to serve— twelve jurors and six alternates. Judge Henderson gave them their marching orders. For the next six weeks, there would be no watching or reading the news, no googling the explosion, no conversations of any kind about the case. When riding the courthouse elevator each day with attorneys and witnesses, the jurors could give a friendly nod, but nothing more.

"You must base your verdict solely on what the parties bring out here in court," the judge said.

On the morning of June 17, a group of lawyers queued up on the vast expanse of pavement outside the courthouse. One by one, they flashed ID, loaded briefcases and suit jackets into the X-ray machine, and stepped through the metal detector. They gathered in groups at the elevator bank, ascending a few at a time to Henderson's court- room, a small, windowless chamber at the end of a hallway on the nineteenth floor.

By 8:30 a.m., the courtroom's wooden benches were packed with spectators. A number of San Bruno firefighters who had responded to the explosion appeared in uniform and stared down the defense. Earley sat among a handful of PG&E representatives. Journalists from the local papers—the *San Francisco Chronicle*, the *San Jose Mercury*

News—prepared to take furious notes. There would be no photos or video cameras. At 9:04 a.m., the jury filed in.

The trial would be the ultimate crash course in the mind-numbing world of pipeline regulation, governed by acronyms and formulas that engineers spend their entire careers memorizing. Both the prosecution and the defense had spent years with the material. But the jury, a diverse group of people including an insurance analyst, a computer technician, and an auto dealership worker, would have to learn a new language. What was a pup? A pig? MAOP? SMYS?

Hoffman opened for the prosecution. Hoffman, at about forty, had been working in the US Attorney's Office for most of her career. She had thick, shoulder-length brown hair and bright brown eyes. That morning, she offered a brief outline of the government's case against the company. On a screen appeared a photo of the destruction in San Bruno.

The prosecution's case, in all of its complexity, boiled down to a few key elements. The Pipeline Safety Act, enacted in 1968, was amended in 2002 to require pipeline operators like PG&E to identify risks along lines that ran beneath populated areas, where an accident could have major consequences. The operators were required to gather records on the line, rank the risks, and assess them appropriately.

Operators were supposed to pay close attention to risks that could compromise the integrity of their pipelines, many of which have welded seams. If any part of a seam is faulty, it's vulnerable to bursting open under pressure. That meant operators were obligated to test those lines to ensure they could withstand high operating pressures.

"If you want to test the integrity of a pipeline—whether or not the pipeline can still safely operate—there are two options," Hoffman explained. The first was a hydrotest, a straightforward means of determining the maximum pressure a pipeline can handle. The test involves draining a line of gas, filling it with highly pressurized water, and then monitoring it for leaks or ruptures. It was inconvenient and messy, but it was the best method of testing seam weld strength.

The second option, known as in-line inspection, involved using tools called smart pigs, digital devices that swim through pipelines and collect data as they go. The pigs could assess a range of integrity issues so long as the pipelines were specially outfitted to accommodate them.

Hoffman tried to make it memorable for the jury. "This tool, as it's going through, kind of makes . . ."

She paused to make a squealing sound.

"So it's called a smart pig," she said.

Both testing methods were more effective, and more expensive, than a third testing option known as "external corrosion direct assessment." As its name suggested, the method involved surveying the outside of a pipeline for corrosion. It could not be used to assess weld strength, something PG&E's engineers knew. But they nevertheless chose direct assessment to inspect pipelines with risky welds, the prosecution alleged, not because it was sufficient, but because it was cheap, and they were under pressure to cut costs.

The Pipeline Safety Act also had rules regarding operating pressure. Pipeline operators were supposed to keep gas pressure at or below the maximum level reached within specific time frames outlined in the new regulations. If a pipeline exceeded that threshold by any amount, the regulations required operators to test the integrity of its seam welds. There was some leeway—but only if the pipeline had been hydrotested to prove it could safely operate at the higher pressure.

Prior to San Bruno, PG&E had more than eighty miles of lines that had experienced pressure spikes above their legal thresholds, and none had already been hydrotested. That meant PG&E was required to prioritize them for seam-weld inspection.

But PG&E didn't do that, the prosecution alleged. PG&E chose to prioritize them only if gas pressure exceeded the legal threshold by more than 10 percent—a deliberate decision to ignore the law's requirements. That led to the problem with the NTSB. PG&E had given investigators a document outlining its 10 percent policy, and then tried to walk it back by saying the document was merely a draft of a policy never implemented. The prosecution alleged this was a lie that impeded the agency's investigation.

"Ladies and gentlemen, motive is not an element of any crime the government must prove to you," Hoffman said. "However, we expect that the evidence is going to show you that as PG&E was cutting its spending in areas that ensured the safe operations of its pipelines, at the very same time it was taking actions to maximize its profits for the corporation."

The energy in the room shifted as the prosecution ceded the floor to the defense. Bauer, a trim man with narrow shoulders and thinning hair cropped close at the temples, looked the part of a corporate defender. He favored dark tailored suits with crisp white shirts. And he knew how to channel authority, making eye contact with his audience while speaking in short, direct sentences. He had pored over the millions of pages of documents drawn from the company's archives and studied their technicalities, determined to outshine the prosecution in his mastery of the material.

"Well, we have heard some harsh words in the opening statement by the government," he said as he took the floor. "Said there was a pattern of criminal conduct; a deliberate choice to make pipelines unsafe. A cover-up. Deliberately misleading an investigation of the explosion. And that the company was cutting costs in order to maximize profits. Now if only the world were that simple. Right?"

That set him up for his own version of the "Where's Waldo" argument. He reminded the jury that a corporation, in and of itself, can't commit a crime. It's only as guilty as its employees.

"It's easy to snarl at a logo," Bauer said. "It's much harder to look a person in the eye and say 'You are a criminal. You are a criminal who made these pipelines unsafe.' And you notice the government didn't do that."

For the next hour and forty-five minutes, Bauer inundated the jurors with pipeline minutiae. It was part history lesson, part tutorial, replete with metaphors to help them grasp complex operating principles. He wanted them to learn the basics. But he also wanted them to recognize the extent of what they didn't know. The upshot was clear: pipeline regulations were not as black and white as the government made them out to be, so the question of whether PG&E violated them was fraught with challenges.

And then the first witness took the stand.

For a two-week stretch in June, the jury heard from a parade of witnesses, first federal pipeline regulators and then PG&E employees. The prosecution flashed evidence of the company's poor record keeping and haphazard approach to managing changes in pipeline pressure. Bauer scored points on cross-examination, homing in on the many confounding variables that muddied the case.

Then, on the morning of July 7, the defense's hope for victory began to fade. The jury filed into the courtroom to find Bill Manegold, a lifelong engineer who had spent thirty-five years at PG&E. He had retired in 2014 after nearly a decade within the gas division supervising transmission operations. He was one of the prosecution's star witnesses.

It wasn't Hoffman who began the examination, but Hartley West, who had worked in the US Attorney's Office for nearly fifteen years. She loved the challenge of absorbing and distilling complex regulatory cases. With Manegold on the stand, West deployed one of the prosecution's most effective weapons: a cache of emails between PG&E gas engineers showing their consternation as their budgets were repeatedly cut during the Business Transformation era.

In 2008, Manegold fielded a note from the manager in charge of the gas transmission budget who earlier flagged the need for urgent cuts. "As expected, we got saddled with a very low 2009 budget," it read. "What was unexpected was how low it was, basically equivalent to 2008."

"We can probably knock the leak investigation budget down to 200k again in '09," Manegold responded.

Already, the gas division's integrity management engineers, in charge of looking for leaks and other problems, had been forced to scrap plans to send smart pigs through a number of pipelines and instead subject them to direct assessment, at a much lower cost.

"Gas Engineering would strongly prefer to smart pig PG&E's higher stress pipelines to obtain a much better initial evaluation of the line, but that is not financially viable at current funding rates," one internal memo read.

West then displayed a PowerPoint slide setting the "ground rules for our discussion" during an off-site meeting for company officers in 2008. It had two categories: "What Is Up For Debate" and "What Is Not Up For Debate," each with several bullet points beneath.

"Do you see that?" West asked Manegold.

"I do," he replied.

"And the first bullet point that is 'not up for debate' is 'eight percent EPS growth,' correct?"

"Correct," he said, affirming that EPS stood for "earnings per share."

"Do you see those listed as 'What Is Up For Debate'?" West asked. "'Safety' and 'Reliability'?"

"Yes," Manegold responded.

It was the sort of document that surely had a backstory. If it meant what it appeared to mean, who could have been so dense as to put it in writing? In the context of the trial, however, it underscored the point that seemed to come up in every other email about budget cuts: PG&E prioritized profits above all else. The jurors were aghast; the journalists in the audience scribbled frantically.

The jury returned the next day to find Manegold again on the stand. At issue was a memo that he began preparing in the fall of 2009, several months before the CPUC was due to audit the integrity management program. Manegold had been nervous. He knew PG&E had operated more than eighty miles of pipelines at pressures above their legal thresholds. The company couldn't produce records showing they'd been hydrotested, so it was obligated to prioritize them for thorough testing. But it hadn't done so, and didn't plan to, prior to the audit.

So Manegold, after consulting with a number of other engineers within the division, took over an effort to draft a memo justifying that decision. Although federal regulations state that "any pressure increase regardless of amount will require assessment," Manegold wrote, PG&E would interpret that to allow a 10 percent breach prior to assessment. Ahead of the audit, he emailed it to several colleagues and consultants, one of whom sounded a note of caution. Federal regulations seemed absolute. But two signed off.

"Bill, sounds reasonable and defendable," one of them wrote.

Manegold responded in jest. "That's what Custer told the lieutenant who said, 'Let's stop here,'" he wrote, referring to Lieutenant Colonel George Custer's devastating defeat in the Battle of Little Bighorn, after his troops camped too close to Native American warriors.

"What did you mean by that?" West asked.

"Like Custer, I thought we were going to get killed in the audit," Manegold responded. "I thought we were going to get slaughtered."

In May 2010, Manegold went into the audit fearful that the regulators would attack his team for flouting federal statute with the 10 percent policy. Fortunately for him, it never came up. When he returned to his office, he shredded the document. By his own admission, he thought the company never should have put something like that on paper.

"Would you agree that you thought PG&E would have been required to assess for manufacturing threats?" West asked, referring to seam-weld tests after pressure increases.

"Without a better justification than what I provided in that memo, yes," Manegold said.

"Why did you shred it?" West asked.

"Because I didn't like it," Manegold replied. "I didn't think it satisfactorily [addressed] the issues that needed to be addressed."

Four months later, part of San Bruno had been destroyed. A seam weld, the sort of thing that required assessment after pressure increases, had failed during an unexpected increase resulting from botched electrical work at a gas control center. Executives were panicking. Manegold exchanged emails with a friend of his in another department after Chris Johns, the president of the utility, stopped by Manegold's office to talk about the company's gas transmission processes. Manegold told Johns that a big part of the problem was the company's old pipelines. A lot of them needed replacement. Even a well-maintained 1956 Cadillac, Manegold told Johns, is not as safe as a new car.

"What I should have said was, in this review process don't assume that the people working on this stuff now are idiots that should be ignored (as it seems to me has been done)," Manegold wrote to his friend.

His friend answered with a long, frustrated response about the gas division's paltry budget, referring to cuts as "money that engineers would have used to add safety margins, replacements, et cetera, over all these years and, yes, maybe not meeting the authorized rate of return," he said. It was a reference to the unusual way that utilities make money—by spending heavily on capital investments, sometimes at the expense of day-to-day maintenance and operations.

"This problem is across gas and electric. I only thought about electric and thought it was just there, but now we see the gas side, too, the piping and valves and design," he went on. "This is the (near criminal) fault of all the financial people running the show since the 1980s, and Peter and Chris continued that same philosophy and, therefore, are not free from fault. The truth needs to be fully aired out to the state and federal authorities."

Who, West asked, was Peter?

Darbee, Manegold replied.

Darbee, as CEO, had been a financial person running the show. Though he might not have realized it, the returns he promised shareholders had come at the expense of pipeline maintenance.

In bringing its witnesses before the jury, the prosecution was not intending to pin the corporation's crimes on any one employee—not Manegold, and certainly not Darbee, who was too far removed from day-to-day pipeline decisions to bear direct culpability for the problems. As Haggarty put it, this was not a crime in which one person pulled the trigger. It was a crime committed slowly, over the course of decades, as corporate strategy shifted to prioritize shareholders, forcing middle-level employees to respond accordingly. The point was to show that a number of them each held information about the risks the company's pipeline system posed and, through their interactions with each other, ignored them or made them worse to the point of breaking the law.

The prosecution began its closing arguments on July 26. The trial had lasted for nearly six weeks. Some days, the jurors sat for even longer than initially contemplated, hearing from witnesses until 3:30 p.m. They were ready to be done, having asked the attorneys a couple of weeks earlier to "commit to being succinct." Jeff Schenk, who, in his midthirties, had already established great skill at trial, delivered the closing argument.

"Pacific Gas and Electric is a company that lost its way. For many years, it forced its engineers to act more like businessmen and businesswomen than engineers. They made decisions to maximize profits instead of prioritizing safety," he told the jury. "This case is about choices. Deliberate choices, and decisions that PG&E made over the years. And really, about choices in three areas. With regard to its records; with regard to threat assessments; and finally, with regard to obstruction."

When Schenk was done reviewing the most damning pieces of evidence in the government's arsenal, Bauer took the floor. It was his last chance to remind the jurors that PG&E was nothing more than its employees, the majority of whom were hardworking people who meant no harm. He lambasted the prosecutors for cherry-picking data and witnesses.

"They chose the people that they thought would make PG&E look the worst," he told the jury. "My question to you: Did you see a group of people who were just so despicable that they were sitting there trying to figure out ways to break the laws for one reason or another, making our pipelines unsafe? Or did you see a collection of people, chosen by the government, who seemed to be decent people, competent engineers, committed to their jobs, doing the best they can under the circumstances, acting in good faith?"

He asked the jury to recall Manegold's explanation for why he shredded the memo he had been worried to present at the audit.

"He didn't say, 'Because I thought it was illegal, and I thought I was breaking the law, and I'm a terrible person, and I'm making pipelines unsafe,'" Bauer said. "He didn't say that at all. He said, 'I wasn't comfortable with it. It wasn't complete. It wasn't a robust enough analysis.'"

That day, Henderson instructed the jury on how to think about the linchpin of the case: the company's willful intent. The defense had argued for a very strict definition—proof beyond a reasonable doubt that an "employee of the defendant acted with an evil-meaning mind" with "knowledge that his or her conduct was unlawful." That would have been a high bar to clear. Henderson saw it differently. Willfulness, he wrote, would be defined in this case as employees' "disregard for the governing regulation and an indifference to the regulation's requirements."

The jury foreperson was a man named Tom Hoffman, a longtime Bay Area resident who lived in Richmond with his wife and two young sons. He had been a juror twice before on simple, short-lived trials, the most memorable of which involved a fight among three women at a local burger joint. Never had he been involved in one of such consequence. Hoffman's boss at the California Academy of Sciences, where he worked at the time as a benefits manager, had wanted him to argue for his dismissal; a PG&E executive sat on the board of the organization. But Hoffman didn't care. His family and friends were all PG&E customers, and he felt a duty to serve. The weight of the responsibility had hit him on the first day of trial.

At first, the jurors were split. The testimony, in its entirety, had been extraordinarily complex. There was no question that gas transmission engineers had been under pressure to cut expenses at a time when the company had committed to earnings growth. Evidence showed they responded by reducing inspection costs, the danger of which was exacerbated by gaps and errors in the company's various records repositories. Less obvious was the matter of willful disregard for and indifference to regulatory requirements. Courthouse staffers stacked the jury room with hundreds of binders and filing boxes stuffed with documents. "Dig in," Hoffman told his fellow jurors. The twelve of them papered the walls with maps and records.

For all the back-and-forth over the challenge of convicting a corporation, Hoffman didn't think it was all that complicated. The jurors had heard testimony from dozens of witnesses, seen hundreds of email exhibits, and read through thousands of pages of records. It wasn't just Manegold who knew that the company wasn't following the rules. Exhibit after exhibit offered evidence that others knew it, too, and failed to correct course. The other jurors came to agree. As they deliberated, they didn't dwell very long on individual employees but rather discussed the body of information contained in the emails and memos outlining the difficult choices they had to make. In Hoffman's mind, it was clear that the company valued profits over safety, a dynamic that resulted in its middle-level managers making decisions that didn't comply with the law.

The jury returned its verdict on August 9, 2016. The company's poor record-keeping practices, it determined, didn't rise to the level of a felony; the prosecution hadn't been able to point to specific employees who had deliberately allowed the problems to accrue over time. But its inspection problems did. The jury found the company guilty on five counts of violating the Pipeline Safety Act, and one count of obstructing the NTSB's investigation.

Haggarty felt a sort of numbness as the jury delivered its verdict. He had, at that point, been working on the case for more than four years. It had consumed his weekends and holidays, and in a moment, it was all over. It took some time before he felt a sense of deep vindication. The prosecutors hadn't won their case in its entirety, but the partial victory was still a major achievement, one that wouldn't have happened if Haggarty hadn't spent hours on end combing through millions of pages of documents. After all, he had been the one to find the evidence supporting the obstruction charge.

A wrinkle in the case offset some of his gratification. During the jury's deliberation, the prosecutors, without explanation, rescinded their call for an alternative fine. They never revealed why, but it likely had to do with the sheer challenge of proving PG&E's financial gain as a result of the crime. Haggarty was incensed by the decision. If they couldn't prosecute individuals, what other choice did they have than to seek the largest possible fine from the company? Earlier in the case, the team had brought in forensic accountants to help arrive at a number. But the prosecutors had their reasons, and stuck to them. At that point, they were tired. The trial had been all-consuming.

Instead of an alternative fine, PG&E would pay no more than the statutory maximum for each of the six charges, for a total of $3 million, a figure that amounted to little more than a rounding error on the company's balance sheet. The paltry amount demonstrated just how difficult it is to penalize a corporate criminal. Corporations are fined just as people are, despite their far greater resources.

In January, Henderson issued his sentence. Just as he did with the Oakland Police Department and the state prison system, he devised a rehabilitation strategy. He put PG&E on probation for five years, during which time it would submit to the oversight of a federal monitor with unfettered access to books and records. The company would have to develop a compliance and ethics program. Employees and executives would have to complete ten thousand hours of community service, mostly in San Bruno. There was a public shaming element, too. Prior to its conviction, PG&E had spent heavily on prime-time TV commercials portraying its innocence in the explosion. So Henderson required them to buy those same spots to explain its crime and the remedial steps it was taking, as well as full-page ads in the *San Francisco Chronicle* and *The Wall Street Journal*.

For all the requirements, there was one rule of probation that PG&E had to follow above all else: no more crimes.

Nine

FIRE SIEGE

I n November 2016, three months after the trial, PG&E announced that Tony Earley would soon step down as CEO. The decision capped a tense few months within the boardroom. Several of the directors thought the company should appeal its conviction. Most of the senior management, including Earley, vehemently disagreed. The fight was over. It was time for the company to move on.

Earley, who had built his life in Detroit, decided it was about time for him to move on, too. He had never quite fit in in San Francisco, surrounded by larger-than-life tech executives and their flashy corporate culture. By comparison, Earley was old school. He sat on the board of Ford Motor Company. Shortly after becoming CEO of PG&E, he flew company executives to General Electric's legendary leadership school in Crotonville, a small town just north of New York City on the Hudson River. Opened in 1956, the sixty-acre campus was the place where GE executives taught up-and-comers the secrets of twentieth-century success. Earley had also taken his subordinates on a field trip to DTE Energy, the utility company where he had spent much of his career. In 2014, during an appearance at the Detroit Economic Club, he assured the audience that he would be home before too long.

"We are coming back to Michigan when this adventure is over," he promised. "One of the reasons is someone once summed up San Francisco absolutely perfectly. They described the city as forty-nine square miles surrounded by reality."

The board had considered two candidates for Earley's replacement: Nick Stavropoulos, who had rehabilitated the gas division after San Bruno, and Geisha Williams, who had risen to become president of the electric division in 2015. The directors ultimately selected Williams. Stavropoulos, meanwhile, was promoted to chief operating officer of the utility.

When Williams took the helm in March 2017, she became the first Latina to ever run a Fortune 500 company. It was the pinnacle of a remarkable career. She came to America at age five, speaking no English, after her father, a political prisoner, emigrated from Cuba with her mother in 1967. She spent her adolescence working as a cashier in her family's grocery store and then enrolled at the University of Miami, becoming the first in her family to graduate from college. Her love for the utility business started with Florida Power & Light.

At first, she hadn't imagined becoming more than an engineering supervisor. But she had a mentor who recognized her potential and pushed her to think bigger. Someone had to run the company. Why not her? Those words stuck with her, pushing her to work exceptionally hard. Many people, women especially, found her inspiring. She was also carefully rehearsed. Climbing the rungs at Florida Power & Light and then PG&E required a certain sort of diplomacy. A California Public Utilities Commission staffer, in a snide email exchange with a PG&E employee, had once referred to Williams as the "senior vice president of bullshittery."

When Williams became CEO of PG&E, the company had made great strides on meeting its clean energy targets. Renewable sources served almost a third of customer demand in 2016. Rates, too, had gone up. That year, PG&E's residential customers paid an inflation-adjusted twenty cents per kilowatt-hour on average, among the most expensive of any large utility in the country. That wasn't lost on Williams. On an earnings call with investors and analysts, she emphasized the company's commitment to affordability. The company, she said, had plans to reduce costs by $300 million in 2017. That involved laying off eight corporate officers, three hundred ninety staff members, and eight hundred contractors. On top of that, five hundred open positions would go unfilled. The company was on a mission to tighten the budget. "You can expect to see more of that coming from us in the years to come," Williams said.

Meanwhile, PG&E could no longer ignore the escalating threat of wildfire. By 2016, the danger foreshadowed by the 2008 fire siege had manifested throughout the company's service territory. Throughout the state, about one hundred million trees had died, destroyed by bark beetles and lack of water. More than half the state faced severe drought conditions. The company would have to do much more to keep its power lines clear of branches, especially when the Diablo winds blew.

The CPUC still hadn't required PG&E to prepare a formal wild-fire plan, after company attorneys successfully argued that Northern California didn't face the same risks as Southern California. Agency staffers, in 2012, had agreed.

But even then, PG&E was becoming aware of mounting fire risk. In 2011, when Williams began overseeing electric operations, the company received a report it had commissioned to determine how many fires were caused by trees touching its power lines. Between 2007 and 2010, during the first major drought in years, 186 fires had started when branches fell or trunks snapped, taking out live wires on their way down. About 40 percent of the fires occurred in the dense forests north of San Francisco, including those in Napa and Sonoma Counties.

In 2014, in the middle of the next major drought, the CPUC directed all three of the state's large utilities to do more to address the threat of fire. PG&E sent out more contractors to trim trees near live wires. The threat was daunting. In a presentation that year, the company acknowledged that California faced year-round wildfire risk as a result of the drought, echoing what Arnold Schwarzenegger had said years earlier as the 2008 fires raged. The company realized there was no way it could clear every tree away from every line, a Sisyphean undertaking. So it attempted to identify the ones at highest risk of causing fires. "It's not about the 50,000,000 *potential* threats . . . it's about the 5,000 *real* threats to our facilities," one slide read.

PG&E faced a significant challenge in ramping up tree trimming. The company relied entirely on contractors to do the work, and there weren't enough of them to go around. It resorted to allowing one contractor to lower its hiring requirements in order to get more people in bucket trucks. That same contractor, which at one point handled nearly half of PG&E's tree-trimming work, had consistent performance problems, missing trees and falling behind schedule. In 2014, PG&E's head of vegetation management emailed the contractor's regional manager to express his disappointment that an ongoing audit showed that crews were failing to meet their tree targets. "Almost complete and not looking good," he wrote, reminding the manager that the crews had failed a similar audit just a couple of years earlier. "I do not want to see a repeat and I'm sure you do not either."

A few years later, Williams, as CEO, would confront a sobering reality. The years-long focus on renewable energy had diverted the attention of both the company and its regulator. As PG&E secured new wind and solar contracts, its service territory became a tinderbox, and its rates reached new highs. Now, it would have to cut employees, contractors, and spending at a time when climate change had come to pose a real and immediate threat. It would have to do more, with fewer resources, to patrol power lines and clear them of trees in a race to get ahead of the problem. But the problem had a substantial head start. PG&E lost as soon as the race began.

The evening of October 8, 2017, was dry enough to invite the wrath of the Diablo winds. The state fire service warned of extreme fire risk throughout Napa and Sonoma Counties, where gusts were expected at anywhere between fifty-five and ninety miles per hour. At

9:41 p.m., a resident of the small town of Calistoga, north of the city of Napa, reported a fire burning bright along a road that separates an expanse of vineyards from heavily forested hills. The fire spread west with devastating speed, fanned by easterly winds racing out to sea.

The winds funneled into a steep ravine northeast of the Sonoma County seat of Santa Rosa, reaching eighty miles per hour as they swept through neighborhoods. The structures trapped the gusts, creating fiery tornadoes strong enough to flip cars, uproot trees, and strip roofs off houses. Electrical transformers exploded. Propane tanks, too. The hills of Santa Rosa looked and sounded like a war zone. Thousands of people hurried to stuff belongings into bags and raced to their vehicles to join a mass exodus.

Will Abrams awoke in the middle of the night to the sound of his two children screaming. The smoke detectors near the back of the house were wailing. At first, he thought it was yet another false alarm—the sensors were easily triggered, sometimes going off when heavy fog rolled in. But the kids knew better. "We're going to die!" his son yelled. He was nine years old. He yelled it again, over and over.

Abrams ran outside. The back of the house, built atop a wooded hill overlooking a vineyard, was on fire. He raced back inside to herd his family out. He took off his Birkenstocks and threw them to his son. His wife and six-year-old daughter hurried through the door. They piled into a Kia sedan and started down the long driveway, a fifty-yard slope lined with burning branches. Abrams reclaimed his sandals and darted from the driver's seat to pull the branches aside. They sped from their cul-de-sac to a winding two-lane road leading out of the burning hills. To the left was nothing but flames. He turned right, pushing ahead through blinding smoke and burning trees. His son had his head in his hands, berating himself for forgetting to save a teddy bear that had been in the family for two generations, its worn seams thick with extra thread. After what felt like hours, Abrams saw a fire truck in a parking lot and pulled over. The firemen pointed them to an evacuation route. They joined a line of cars inching south toward San Francisco and arrived at Abrams's sister-in-law's house in Oakland in early morning darkness.

That fire was one of more than a dozen that ignited overnight that Sunday as wind-whipped embers showered dry forests throughout Northern California. Many of them spread through wine country, torching homes and vineyards. Others ignited in the Sierra foothills and the Mendocino National Forest. By Wednesday, twenty-three people were dead. Tens of thousands of acres were burning as fire crews struggled to contain the spread. Entire neighborhoods had been razed. More than sixty thousand people had been evacuated, with thousands more continuing to flee.

Abrams knew there would be nothing left when he returned to Santa Rosa, but he had allowed himself a flicker of hope that perhaps the laws of nature hadn't held. A photograph from a neighbor snuffed it out. The house was destroyed, windows blown out, the yard a mess of metal and debris. Some of the beams stood, but everything inside had burned. He and his family returned two weeks after fleeing to dig through the ashes. Somehow, his father-in-law's wedding ring, stored in the garage after his mother-in-law passed away, survived intact. So did a metal Band-Aid box full of old coins Abrams's parents had collected on trips abroad. There was almost nothing else. The bear didn't make it.

More fires started throughout the month. Southern California also suffered severe devastation. But Northern California, where forty-four people died and more than six thousand six hundred homes burned, saw the worst of the destruction. With the fires still burning, Williams went to PG&E's operations base in Sonoma County. She hadn't fathomed the extent of the destruction. Houses had collapsed onto their foundations, reduced to piles of metal and cement. Cars had been stripped of rubber and glass, their scorched metal frames like ashen cages lining suburban streets. Blackened tree trunks stood erect in the smoky air, leaves and branches gone. At the start of November, when the company reported quarterly earnings, Williams told analysts that the call would be different than usual. "I've seen firsthand the destruction that hurricane force winds can bring to a community, but this was like nothing I've ever seen," she said.

The financial implications were grim. At issue was inverse condemnation, the California constitutional provision that allows residents and business owners to seek compensation from utilities if power lines somehow damage their property. If PG&E's power lines were involved in any of the fires, it would be on the hook to pay damages. "Even if a utility has followed all the rules and, in essence, has not done anything wrong," Williams told analysts on the call.

The CPUC, meanwhile, was at last preparing to hand down its decision on the Witch Fire that had blazed through the San Diego foothills a decade earlier, foreshadowing crushing liability for utilities. At the end of November, the CPUC ruled that San Diego Gas & Electric would not be allowed to recoup $379 million in fire-related damages by raising rates. The agency determined that the utility had failed to operate its electric system in a "reasonable and prudent" manner and should therefore have to eat the costs. The precedent had huge consequences for all three of the state's largest utilities—especially PG&E.

State fire investigators still hadn't determined whether PG&E's equipment had sparked any of the fires that had swept its service territory, the most destructive of which were known collectively as the North Bay fires. But PG&E had filed reports notifying regulators that it had found damaged power line equipment near the suspected ignition points of more than a dozen of them. On December 20, the company announced it was suspending its dividend to conserve cash. Its shares fell 13 percent.

The lawsuits were already trickling in. Trial lawyers throughout the state were finding victims and filing claims on their behalf. The company would soon face more than a hundred lawsuits seeking compensation for fire-damaged properties. It wasn't just single-family homes, though there were thousands of them. The fires had also ripped through the heart of one of California's most valuable industries. Vineyards and wineries worth millions had been destroyed.

In August 2017, five months after Williams became CEO, Judge Thelton Henderson retired. He would no longer be the one to oversee the company's five-year probation following the San Bruno trial. Judge William Alsup, a spry seventy-two-year-old with neat white hair and square glasses, inherited the case. Alsup, born in Jackson, Mississippi, had been nominated to the federal bench by President Bill Clinton in 1999 after more than two decades of back-and-forth between private practice and roles within the Department of Justice. He had an unorthodox approach to learning his cases. He loved asking both sides for long and detailed explainers on the facts underlying their arguments. Those often led to more questions, and more requests. Initially, he didn't ask much of PG&E—his first meeting with company attorneys lasted just eleven minutes. It wouldn't be that way for long.

In December, two months after the fires, the attorneys again met with Alsup for the company's regular probation check-in. Alsup, an avid hiker whose own photographs of the California wilderness hung in the federal courthouse, was painfully aware of the destruction wrought by the fires. He seemed puzzled about the scope of the probation conditions, one of which required PG&E to submit to a monitor overseeing its gas operations. Alsup wondered whether that oversight extended to include power line operations. In court, he addressed the monitor, an attorney with Kirkland & Ellis.

"So your work," he said. "Is it limited to gas, or do you also get off into electricity?"

"The technical requirements are focused on gas operations," the monitor replied, adding that the company's commitment to safety and compliance was supposed to span the gas and electric divisions.

"Do you have any responsibility or jurisdiction over inquiry into the North Bay fires, as to what went wrong there, if anything, by PG&E?" the judge asked.

"Your Honor," the monitor replied, "I think responsibility and jurisdiction is a question for you."

The California State Legislature began its session in January. The lawmakers were in a frenzy. Immediately after holding a moment of silence for those who died, several senators from fire-ravaged districts introduced a bill that would codify the CPUC's recent ruling by prohibiting the state's utilities from seeking to recover fire liability costs from customers if the companies were found to have acted "unreasonably" in maintaining their power lines. PG&E swiftly came out against it, arguing that allowing the state's utilities to face "essentially unlimited liability" would sap their financial health and impede their ability to sell shares and issue debt to investors.

"At a time when California is asking privately owned utilities to invest billions of dollars to meet the state's greenhouse gas reduction goals, these risks pose real consequences for the state's environment, economy and communities," PG&E said.

At the end of January, Williams flew to New York City to deliver the keynote at the United Nations Investor Summit on Climate Risk, where Peter Darbee, a decade earlier, had warned about nearsighted thinking. PG&E, Williams said, had derived a third of its electricity from renewables in 2017, three years ahead of the state-mandated deadline. She cautioned that the potential for massive fire-related liabilities threatened to hinder further progress. The fires, she said, "exposed a hidden and terrible irony."

"Climate change doesn't just pose a threat to life and property," she said. "It also has the potential to disrupt climate action. We need to think about this broadly as a society. We're going to have to look at climate change and its consequences differently."

That month, the CPUC at last finished mapping areas at high risk of fire, nearly a decade after first starting to mark them in yellow and red. The final product showed that almost all of Northern California was yellow, save for the Central Valley. The Sierra foothills appeared as lines of red, as did the wine country region and the hills

surrounding San Francisco. To the south, the areas surrounding Los
Angeles and San Diego were almost entirely red.

Meanwhile, PG&E had been reporting more fire incidents each
year. Since it began recording fires in 2014, its equipment had started
about 1,550 them. The numbers were up sharply in 2017, a particularly bad year, with more than 500 fires igniting throughout its service territory. Most of them were insignificant. But the potential for
their spread had never been greater.

In March, PG&E announced a sweeping risk mitigation plan. It
had nearly completed a new operations center dedicated to measuring fire risk around the clock. It planned to install more weather
stations and cameras to monitor conditions on the ground. And it
said it would consider something drastic: proactively shutting off its
power lines when hot, dry winds picked up. No power, no spark.

Come June, Cal Fire had released some of the results of its 2017
wildfire investigations. It determined that PG&E's power lines had
sparked at least sixteen of them. The company issued an emergency
notice to investors. It planned to take a $2.5 billion charge in antici-
pation of steep losses. That was more than the $1.66 billion in profits
the company had earned in 2017. And the company cautioned that
the charge was conservative—total damages would likely well exceed that number. Analysts dialed in for a special call with executives. Rumors were flying that the company would have to seek
bankruptcy protection. Williams didn't rule it out. But she said the
company was working with lawmakers to try to figure out a way to
avoid it.

"You have to realize that many of the lawmakers here in California vividly remember the energy crisis and what came as a result of
that," she said.

At that point, the company faced more than two hundred civil lawsuits. Individuals had filed them, as had cities and government agencies. Insurance companies were also seeking reimbursement for claims they had paid to homeowners and businesses. Trial lawyers began summoning dozens of PG&E employees for depositions. Williams sat for hers in July. One of the most dogged attorneys suing the company began the interrogation. PG&E had clearly accepted some level of risk when it came to wildfires. He wanted to know exactly what that threshold was.

"We aspire to have absolutely no wildfires," Williams said. "But at the same time . . . when you've got over a hundred thousand miles of conductor running through, many times, national forests, state forests, and everything else, there is an exposure—an inherent exposure that comes from running an electric utility."

Around the same time, Nick Stavropoulos announced he would retire from his role as chief operating officer of the utility in the fall. He told his colleagues that they needed a leader with the sort of energy he had brought in overhauling the gas division, and he simply didn't have it anymore. The electric division was only just beginning to assess its weaknesses. He had been bitterly disappointed when the board chose Williams, instead of him, to fill the top spot. But he had come to realize it was a blessing in disguise. He planned to be back in Boston by the end of September.

PG&E spent much of 2018 on a tree-trimming tear. After the 2017 fires, the CPUC had stepped up requirements for power line clearances, requiring the utilities to keep trees at least four feet away from live wires in areas at high risk of fire. Prior to that, a foot and a half had often been acceptable. That meant the company had tens of thousands of trees to clear, and tens of thousands more to re-work to a new standard. By then, more than 129 million trees were dead.

Autumn came, warm and dry. In August and September, a spate of wildfires burned throughout state, fanned by the Santa Anas in the south and the Diablo winds in the north. One of them, an enor-mous fire that scorched more than 450,000 acres in and around the Mendocino National Forest, was the sum of two vegetation fires that merged as they raced through the trees. Crews took three months to contain it. PG&E wasn't implicated in its ignition. But it was yet another reminder of how risky its territory had be-come.

State lawmakers, nearing the end of their legislative session, rushed to hammer out the details in a suite of bills dedicating state resources to address the crisis. One of the most significant measures lessened the sting of inverse condemnation by allowing PG&E to issue low-interest bonds to pay some of the liability costs arising from the 2017 fires. The company would repay the debt through a surcharge on customer bills. PG&E estimated that an average resi-dential customer would pay an extra five dollars a year for every $1 billion in bond debt. It was a bid to keep the company out of bank-ruptcy.

The legislation also addressed future risk. Utilities liable for fires igniting in 2019 and beyond would be allowed to recoup some of those costs through customers, so long as regulators determined that the company hadn't been negligent. But the bill said nothing at all about fires starting in 2018.

In October, a year after the fire siege, PG&E was determined not to have another. It had warned months earlier that it might proactively shut off its power lines to reduce the risk of sparks when the Diablo winds returned. That meant cutting power to any number of small distribution lines vulnerable to failure. Such a blackout was dubbed a "public safety power shutoff."

It wasn't a new idea. San Diego Gas & Electric had been the first California utility to consider it in light of the 2007 Witch Fire. It first pulled the plug in 2013 with a blackout that affected a few thousand customers for a few hours. Southern California Edison had plans to do it, too, and had warned its customers that fall that the winds might warrant widespread shutoffs.

The time came for PG&E on October 13, a Saturday. Its wildfire operations center, the new facility opened a few months earlier, was packed with employees from every relevant division—planning, operations, logistics, communications, and government relations. They looked at the wind forecast and decided that a shutoff was necessary. Calls went out to ninety-seven thousand customers in twelve counties, warning them to prepare.

At dinnertime on Sunday, October 14, the lights went out and stayed out. The small town of Calistoga was especially hard hit. The Calistoga Inn, an eighteen-room hotel with a restaurant and brewery, was in the middle of the dinner rush with 150 people dining on the patio. The local hospital postponed surgeries. All three Calistoga schools closed. At the Calistoga Roastery, a refrigerator full of groceries went to waste. Hotels and restaurants canceled reservations at the height of tourism season. Cal Mart, the local grocery store, closed for about eighteen hours, costing the owner thousands of dollars in inventory. It took two days to restore power. All told, PG&E ended up cutting power to sixty thousand customers in seven counties. Many of them were livid. But the company managed to make it through October without having started a fire.

On November 8, 2018, the luck ran out when a hook holding up an insulator string on the old Caribou-Palermo transmission line broke in half, dropping a high-voltage wire. It sparked as it swung in the wind. A proactive shutoff wouldn't have helped—at the time, the company was targeting only lower-voltage lines feeding homes and businesses. Within an hour, one of the largest and fastest wildfires in California history was roaring toward Paradise.

It was scarcely dawn in Hawaii when Kirk Trostle's phone rang. His vacation ended the moment he picked up. It was his daughter, calling to tell him that their town would soon no longer exist. She was frantic, trying to figure out how to rescue Trostle's horses from his property and evacuate with her husband and three sons. Trostle's son was doing the same, trying to get his wife and two sons to safety. As she hung up, she warned it might be the last goodbye.

Trostle and his wife, Patty, could barely process what was happening, overwhelmed by the sort of emotion that eludes description. It was an hour and a half before their daughter called back to say the family had made it out alive. Never had ninety minutes felt so long. They had thought for sure they'd lost her. They spent the rest of the day making phone call after phone call, trying to help their chil-

dren find a place to stay. There were no flights out that afternoon, so
they endured another night in Kauai. They touched down in Sacramento the next day and picked up a few items of clothing. It would
be weeks before they could see if any of their possessions had survived the fire.

In those weeks, the Trostle family lived like nomads. Kirk and
Patty stayed with family in Redding while each of their children
bounced from house to house, unable to camp in any one place for
long. Their daughter finally found a place near Lake Almanor, the
reservoir Great Western had built a century earlier. All of the local
hotels were filled with Paradise refugees. The town was cordoned
off. Fire crews were sifting through toxic debris in search of bodies.

Visits were permitted a few weeks later. Lines of cars formed
as hundreds of anguished residents braced themselves for what
they would find. Trostle and his wife returned to find nothing left of
their home of nearly twenty years. The barn, the swing set for the
grandkids—gone.

Hours after the fire started, PG&E filed a short report with the
CPUC noting that it had recorded an outage on the Caribou-Palermo
at 6:15 a.m. The company disclosed that it had discovered damage to
the line during a midday aerial patrol. "This information is preliminary," the report said. Investors read between the lines. By the end
of the following week, shares in PG&E had fallen by nearly half,
their steepest plunge since the company filed for bankruptcy protection after the California energy crisis. Its stock was in free fall for
five days straight.

PG&E told investors that it had exhausted its lines of credit and warned that its $1.4 billion of insurance coverage might not be enough to cover claims against the company if it were found liable for the fire. Analysts estimated that the company's total liability, including the 2017 fires, could reach $30 billion. That was nearly triple its market value, which had fallen to about $9 billion during the sell-off. In mid-October, it had been $25 billion.

At the end of November, Alsup typed up an order and filed it to the docket. Like everyone else, he could see that PG&E was going to have to answer for the Camp Fire.

"What requirements of the judgment herein, including the requirement against further federal, state, or local crimes, might be implicated were any wildfire started by reckless operation or maintenance of PG&E power lines?" he asked. He gave the company until the end of the year to respond. He invited the federal prosecutors to weigh in, too.

On New Year's Eve, the day that Marc Noel, Butte County's deputy district attorney, boarded a helicopter and flew the Caribou-Palermo in search of criminal evidence, PG&E filed its reply. Its attorneys stated the obvious in a vague, circular response.

"If it were determined that a wildfire had been started by reckless operation or maintenance of PG&E power lines, that would, if the specific circumstances gave rise to a violation of federal, state, or local statutes, implicate the requirements of Special Condition of Probation #1 of the judgment, which provides that while on probation, PG&E shall not commit another federal, state, or local crime," they wrote.

Hallie Hoffman and Jeff Schenk, two of the prosecutors who had led the San Bruno trial, responded on behalf of the United States. They were quick to say that any judgment was premature. But they explained the potential charges PG&E could face if it were found to have acted with criminal intent. San Bruno had exposed the company's "knowing and willful" violation of federal pipeline laws. Perhaps electric division employees had acted with that same state of mind.

"At criminal negligence, PG&E could have committed involuntary manslaughter," the prosecutors wrote. "And at malice, PG&E could have committed murder."

Ten

CHAPTER 22

On January 14, 2019, PG&E announced it would seek bankruptcy protection before the end of the month. Geisha Williams had resigned as CEO just hours earlier under pressure from the board of directors, which had decided she needed to go. The company was in complete turmoil, with electric division executives stepping down one after the other. Its share price hit a new low—$6.36 a share, down from $70 at its peak just before the 2017 fires.

On January 29, the lawyers filed the papers. PG&E would start the Chapter 11 process again, nearly fifteen years after wrapping up its first bankruptcy. Observers snidely dubbed it "Chapter 22." By that point, California fire investigators had determined that the company's power lines sparked a total of seventeen wildfires in October 2017 that killed twenty-two people and burned enough acreage to cover San Francisco six times over. Thousands of individuals and businesses had filed more than seven hundred lawsuits against the company.

It would have been difficult to overstate the significance of PG&E's decision to seek bankruptcy protection a second time. Once again, it threatened far-reaching consequences for the entire state of California. And this time, the stakes were even higher. Compensation for fire victims hung in the balance. So did wind and solar contracts valued at billions of dollars as the company parsed which financial agreements to keep and which to attempt to cancel or renegotiate. Even California's other large utilities risked losing their investment-grade credit ratings on fears that they, too, could wind up in the same position as PG&E if—or when—their power lines ignited large fires. Though their efforts were at times misdirected, California regulators and policymakers had spent years focused on slowing climate change, addressing wildfire risk, and making utility operations safer. PG&E's second bankruptcy jeopardized much of the progress.

Investigators had cleared PG&E in their probe of the Tubbs Fire, the blaze that raced through Santa Rosa, destroying Will Abrams's home and so many others. They determined it had been ignited by privately owned electrical equipment, rather than one of the company's power lines. But lawyers for the victims had collected evidence to the contrary. They would soon push for a jury trial.

The Camp Fire investigation was still underway, but company executives foresaw the results. Already, PG&E faced dozens of lawsuits filed on behalf of at least two thousand people and businesses affected by the fire, with more appearing each week. Bankruptcy committees would soon form to represent various groups of claimants. Fire victims, insurance claim holders, and various governmental agencies would each get their own, led by lawyers who would fight for maximum compensation for their clients.

PG&E told investors that the Camp Fire, combined with the 2017 fires, would likely result in more than $30 billion in liability costs, affirming the dire estimate that analysts had earlier calculated. The number was almost unfathomable. Producing that sort of compensation would require PG&E to somehow raise billions of dollars at a time when its future value had never looked more uncertain. Pension funds and other long-term investors would shy away from the risk, leaving only high-stakes gamblers shrewd enough to win dicey bets. Already, hedge funds and other aggressive investors were buying debt and equity in the company, with the intent to introduce their own plans for its restructuring. It was going to be one of the most challenging and cutthroat reorganizations the utility industry had ever seen.

Such complex restructuring cases typically take years to resolve. But Gavin Newsom, a longtime politician with thick, pewter-colored hair and a blinding smile, didn't have the luxury of time. Newsom had replaced Jerry Brown as governor just a few days before PG&E announced its bankruptcy plans. Having served as Brown's lieutenant since 2010 and mayor of San Francisco prior to that, Newsom had come to understand the company's problems as a regional issue, with implications for all of Northern California. Suddenly, its bankruptcy threatened the entire state. The first thing Newsom did was to appoint a "strike team" to evaluate how the lawmakers could contain the fallout without appearing to bail out PG&E. He wouldn't allow the company to stay in court for long.

The bankruptcy case was assigned to Dennis Montali, a top-tier federal judge with a nasally voice and a silver handlebar mustache. Montali was a formidable arbiter in complicated restructurings, having spent much of his career as a bankruptcy lawyer before his appointment to the bench in 1993. Montali, who loved the challenge of intricate cases, had overseen PG&E's first bankruptcy, which at the time had been the biggest one in his career. The second one would be exponentially more difficult. He would, at one point, call the case the most complex utility bankruptcy in American history. And, under pressure from Newsom, he would have to resolve it quickly.

"Welcome to San Francisco bankruptcy court," he said as he appeared to preside over PG&E's first hearing, held in the federal courthouse a mere fourteen hours after the company filed its papers. "If you don't know, I'm Judge Dennis Montali." Everyone remembered.

Three floors above Montali's courtroom, another judge was less than pleased about the prospect of having a recidivist on his hands. William Alsup had been thinking about what the court-appointed monitor had said just after the 2017 fires—he, the judge, had the responsibility and jurisdiction to do something about them. So, a couple of weeks after PG&E filed its New Year's Eve response to his question about reckless power line operations, Alsup proposed a new set of probation conditions. Why, he asked, shouldn't the company be required to inspect every single part of the electric grid and remove every tree within striking distance of its lines? And why shouldn't it be required, during windstorms, to shut off any and all lines it couldn't be certain were safe? He gave the company two weeks to respond.

"These conditions of probation are intended to reduce to zero the number of wildfires caused by PG&E," Alsup wrote. "This will likely mean having to interrupt service during high-wind events (and possibly at other times) but that inconvenience, irritating as it will be, will pale by comparison to the death and destruction that otherwise might result from PG&E-inflicted wildfires."

PG&E's attorneys filed a lengthy answer about why the company considered the conditions to be not just unreasonable, but impossible. Such requirements would force PG&E to find 650,000 full-time employees to clear some one hundred million trees at a cost between $75 billion and $100 billion, the attorneys estimated. They went on to explain that power shutoffs should be used only as a measure of last resort because they create serious safety risks by cutting electricity to critical infrastructure, as well as people who rely on it for medical reasons. And a widespread disconnection of various transmission lines, they added, had the potential to "destabilize large parts of the Western United States and Canada," a disruption akin to the Northeast blackout of 2003.

Parts of the response were wholly exaggerated (a future chief executive would later estimate that there were eight million trees within striking distance of the company's lines). But the essence of it was true. No utility, much less one the size of PG&E, had the time, money, and labor needed to clear every tree away from every power line and keep it that way indefinitely. And power shutoffs, used indiscriminately, had the potential to wreak havoc. Critical infrastructure, from hospitals to water processing facilities, require electricity. So do cell phone towers, traffic lights, and emergency response networks. Electricity keeps medication refrigerated, charges wheelchairs, runs ventilators, and opens garage doors when evacuations are needed. Long outages would jeopardize all of that. And widespread ones could threaten the stability of the grid.

Alsup summoned PG&E to court at the end of January, the day after its first hearing with Judge Montali. The company had assembled a new defense team with lawyers from different firms specializing in high-profile litigation. The nineteenth-floor courtroom was packed. Journalists had to sit in an overflow room and watch the proceedings onscreen. Alsup, in his black robes, took his seat at the dais, white brows furrowed. He was feeling testy, in no mood for corporate platitudes. The first thing he did was to remind everyone in the room that eight people had "burned alive" during the San Bruno explosion. Now, nearly two dozen more had "burned alive in their cars and homes."

"The tinder dryness of the chaparral in California, that's all it takes to start a fire. High wind. Trees. Power lines. Seventeen wildfires. Twenty-two people killed. One month. October," the judge spat. "So that raises the question, what do we do? Does the judge just turn a blind eye and say, 'PG&E, continue your business as usual. Kill more people by starting more fires'?"

PG&E's attorneys were silent. Alsup sighed.

"I know it's not quite that simple because we've got to have electricity in this state, but can't we have electricity that is delivered safely in this state?"

Then Alsup asked a fundamental question. Why couldn't PG&E make sure the risk of wildfire was zero?

One of the attorneys took the podium facing the judge. He had prepared a lawyerly response. "Bringing the risk to zero is an incredibly complicated series of policy decisions that have to factor in reliability, cost and safety," he said. "There's a tremendous amount of analysis that goes into how best to, for instance, make vegetation management decisions and how aggressive vegetation management should be versus the cost of—"

Alsup cut the lawyer off. He had done some math. By his calculation, the company had paid $4.5 billion in dividends to investors in the last five years. (He was almost exactly right—the company had

distributed about $4.4 billion in dividends since 2013, before sus-
pending them after the 2017 fires.)

"I hear it all the time, 'Safety, safety, safety!'" the judge said. "But
it's not really true. Safety is not your number one thing. You could
have spent more money to cut those trees."

The lawyer retreated in a fluster. He ceded the floor to Kevin
Orsini, the company's lead defense attorney. Orsini, a tall, stocky
man with broad shoulders and wavy hair just starting to go gray, was
from New York. He was only about forty, having made partner at
Cravath, Swaine & Moore at thirty-two. He had been admitted to
the case just a few days before his appearance in court, but it was
clear he had been studying up for weeks, if not months, ready to talk
about specific types of trees in the California chaparral, regulatory
clearance requirements, and shortages of arborists all across the nation. His tactic was to be gratuitously deferential while still toeing
the company line, saying things like "Your Honor is dead right" and
then going on to explain why the judge was dead wrong.

"We readily accept the proposition that something more needs to
be done to address these issues," he said as he took the podium. He
began walking through the many complexities of vegetation management: high-risk trees versus low-risk trees, growth between inspections, dead branches taking flight during windstorms. On top of
all that, he said, many landowners were unwilling to part with trees
on their properties. Orsini planted the idea of the militant home-
steader.

"I've gone out there with some of the crews," he said. "It's an inherently dangerous job not just because you're climbing trees, but
because of people armed with firearms—"

"If someone tried to keep you off with a firearm, then there's the
sheriff," the judge snapped. "Come on. Come on!"

Orsini had gone a step too far, but he did have a point: PG&E faced real opposition from homeowners in removing potentially dangerous trees, yet another reason why the risk of fire would never be zero. Alsup, upon considering all the challenges, decided to hold off on imposing the new probation conditions. He instead asked PG&E to provide even more information about tree-trimming and clearance requirements. The company filed hundreds of pages in response. As the judge considered what he ought to do, he ruminated on how much the company had paid in dividends. Between 2014 and 2017, as investors reaped billions of dollars, PG&E's equipment started some 1,500 fires. More than a dozen of them had been catastrophic. Twenty-four people died.

—

At the start of February, PG&E finally submitted a wildfire-mitigation plan to the California Public Utilities Commission, a decade after the agency first considered whether it should have to file one. The plan proposed an enormous amount of work. Never had the company attempted such a volume in any given year. It said it would trim or remove 375,000 trees and inspect more than 700,000 distribution poles and transmission towers in areas at high risk of fire, tripling or even quadrupling what it had done the prior year.

But the company also gave an implicit warning that all the work wouldn't be enough to avert crisis come fall. It said it would dramatically expand its use of the so-called public safety power shutoff, the fire-prevention tactic its attorneys told Judge Alsup should be used only as a last resort. In 2018, the company had targeted about 7,500 circuit miles of power lines for potential shutoffs if strong winds picked up. Those lines served about 570,000 people. The new program would target more than 30,000 circuit miles. That meant 5.4 million people—a third of its customers—were at risk of having their power shut off when the winds began to blow, possibly for as long as five days. The plan was radical in its scope, a tacit admission that the company could no longer deliver electricity both safely and reliably all the time.

A few weeks later, PG&E reported earnings. It had lost nearly $7 billion in 2018. The company recorded a $10.5 billion charge in anticipation of being found responsible for the Camp Fire, plus an additional billion for the 2017 fires. All told, it had taken $14 billion in fire-related charges. And it was still anticipating liability costs in excess of $30 billion. At that point, nearly a thousand lawsuits had been filed against the company.

Together, the wildfire plan and the earnings report laid bare PG&E's dire circumstances. The company needed to invest billions of dollars to make the grid safer and keep the lights on at a time when it was bleeding money. PG&E, for many years, had failed to strike the right balance between private interests and public safety. That balancing act was now more critical than ever. It couldn't simply ignore investors as it worked to compensate victims and spend more on safety—it didn't have enough money to dig itself out of the hole. That meant it would have to take money from hedge funds.

For all their notoriety, hedge funds play an important role when companies lose value and need to raise a lot of capital in order to turn things around. Hedge fund investors are often the only ones who will front the money and assume the risk. But it comes at a cost. Hedge fund gambles are strategic. They stake their bets only on companies they can push to restructure in a way that inures to their benefit.

For the first time in PG&E's history, hedge funds known for

betting on distressed companies had acquired more than half of the company's outstanding shares from retirement fund managers and other risk-averse investors. During its previous bankruptcy, hedge fund ownership had peaked at about 15 percent and fell quickly after the company completed its restructuring. The new shareholders were crafting a plan that would preserve the value of their equity, a rare outcome in the bankruptcy process.

Hedge funds were also buying up the company's debt. The most influential among the new bondholders was Elliott Management, one of the oldest and most aggressive funds in the industry. The bondholders, too, would soon push their own restructuring plan, one that would protect their investments by raising more equity at the expense of existing shareholders.

Another hedge fund, Baupost Group, was devising a particularly shrewd strategy. It had bought a substantial stake in the company. And it had acquired something else: more than $2.5 billion worth of insurance claims. Insurance companies that paid fire claims to homeowners had the right to seek reimbursement from PG&E, given that inverse condemnation held the company liable for property damage. A number of companies sold their claims to investors for pennies on the dollar instead of waiting on a bankruptcy settlement. Baupost had become one of the most influential players among the insurance claimants, with sway over when—and how—they got paid.

It took Judge Alsup about a month to come up with his solution to the tree-trimming problem. Rather than impose new requirements, he decided to simply turn the onus on PG&E. The company would be required, under the terms of its probation, to meet its own tree targets and comply with state law in carrying out the work. The kicker: it wouldn't be allowed to pay dividends unless it was in full compliance.

The proposal was largely unprecedented in the small world of corporate probation. In setting terms for when the company could reward its shareholders, the judge had essentially given himself control over a decision usually left to the board of directors. Judge Thelton Henderson had established court oversight as the company determined how to improve its operations. Alsup was attempting to take that several steps further, by allowing the court itself to determine what changes were needed.

On one level, his solution made sense. If the company couldn't fulfill its basic safety commitments, why should investors benefit? But in attempting to address one set of risks, the proposal introduced another. PG&E's financial health after bankruptcy would depend on its ability to make regular dividend payments, a significant factor in attracting more patient investors that, unlike hedge funds, would stick around long enough for the company to have a chance to stabilize. PG&E hadn't paid dividends since late 2017, and its share price had suffered as a result. Even if the company fell out of compliance on tree work, dividends would help it raise money to make other safety investments.

Despite the potential consequences, PG&E didn't do much to block the proposal. The dividend, at that point, was a distant concern, as the company had no plans to restore it during bankruptcy. In their response to the judge, PG&E's attorneys simply reserved the right to push back on the requirement later. Such a limitation on the company's discretion to pay dividends, they wrote, "could have significant negative long-term effects on PG&E's ability to reorganize and continue to make necessary capital improvements." In other words, it would eventually need to reward its shareholders in order to reduce wildfire risk and avoid what would amount to Chapter 33.

Alsup called the company back into court in early April. He once again opened the hearing by reminding everyone in the courtroom why they were there. He marveled at the destruction of the 2017 and 2018 fires, which had collectively burned 3 percent of California's acreage. PG&E's power lines started many of the most destructive ones.

The judge called one of the company's attorneys to the podium. The lawyer kept his remarks brief. He said the company mostly accepted the new conditions, but warned that perfect adherence to the law was fundamentally impossible. "There are a lot of trees out there and we don't have eyes on all of them at every minute," he said. He asked the judge to allow the CPUC, rather than the court, to determine its compliance. The judge shook his head.

"It's going to be your old federal judge," he said. "I am capable of deciding whether or not the law has been violated."

PG&E, meanwhile, had been working since January to find a new CEO to replace Williams. It needed a thick-skinned executive willing to forgo the comfort of relative obscurity to lead the nation's most notorious utility through a thorny and emotional restructuring. Few were interested.

Bill Johnson had been among the only ones to seriously consider the thankless job. When he got the call from PG&E, he was head of the Tennessee Valley Authority, a federally owned power producer that supplies electricity to companies serving about 10 million customers in parts of seven Southern states. Just a few days after the Camp Fire, he had announced plans to retire the following year. He was sixty-five. He enjoyed gardening and baking bread. But those things would have to wait. He agreed to one final act beneath an unforgiving spotlight.

At six feet, four inches, Johnson was an imposing figure with quiet intensity and a soft, vaguely Southern-sounding accent. An industry veteran, he had spent years in utility board rooms and tried to keep an even keel. By his own assessment, he was relatively laid back. He was a longtime fan of the Grateful Dead.

Johnson had learned how to weather a media frenzy a number of years earlier, while CEO of a North Carolina utility called Progress Energy. When Progress merged with its neighbor, Duke Energy, in 2012, Johnson became CEO of the combined company. He held the title for less than a day. The company's board abruptly ousted him the morning he assumed the job and put Duke's old CEO in his place. Johnson left with an exit package valued at $44 million, mostly consisting of stock grants acquired at Progress. The jarring experience forced him to reflect on how he handled life's curveballs. He resolved to prepare for them.

In March 2019, Johnson left his spacious condo in downtown Knoxville and headed west to San Francisco. Unlike his predecessors, he would answer to a new board of directors. The old guard had remained mostly unchanged for years even as PG&E cycled through executives. The company had vowed to replace the majority of the board in the wake of the Camp Fire as politicians and regulators demanded changes to its governance.

They weren't the only ones demanding change. Three hedge funds, Abrams Capital Management, Knighthead Capital Management, and Redwood Capital Management, had together acquired nearly 10 percent of the company's shares, giving them enough influence to put forth their own slate of board candidates. Many of them hailed from the world of financial restructuring. Word soon got out that the new board would have heavy representation from Wall Street.

At 8:05 a.m. on March 11, as the company was finalizing its new slate, Michael Picker, who had been president of the CPUC since 2015, put in a phone call to PG&E's interim CEO. He warned the executive that it would be "simply unacceptable" if the company didn't choose directors with the wherewithal to turn around an institution so deeply embedded with risk. He said the company should only consider candidates who believed that they, personally, had the means to address its many problems.

Governor Newsom was also deeply concerned. At the end of March, he sent a strongly worded letter to the company:

> I am troubled to learn that PG&E is primed to reconstitute its board with hedge fund financiers, out-of-state executives and others with little or no experience in California and inadequate expertise in utility operations, regulation and safety.
>
> With this move, PG&E would send a clear message that it is prioritizing quick profits for Wall Street over public safety and reliable and affordable energy service. . . .
>   . . . I strongly urge you to reconsider.

The letter had little effect. A few days later, the company announced it would move forward with its slate. Five of the ten appointees worked or had worked at various investment firms, mainly hedge funds with histories of buying stakes in distressed companies.

Three members of the old board would stay on for a total of thirteen directors. PG&E was a utility company, but it was also a publicly traded corporation, and in the depths of bankruptcy, money spoke louder than either Picker or Newsom.

Shareholders elected the new board at PG&E's annual meeting on a gray, chilly morning in mid-June. It was a somber affair. Days earlier, state investigators had released their final determination on the cause of the Camp Fire, confirming that PG&E's transmission line had started it. Protesters gathered outside company headquarters, where its investors and directors had assembled. On the cold concrete, the protesters laid a pair of shoes for every person killed by the fire. They planted flowers in each one as they read the names of the dead.

PG&E's dysfunction weighed heavily on Michael Picker. He had for months been grappling with the fact that something was fundamentally wrong with the way the company operated. A few weeks after the Camp Fire, he had opened a new proceeding to determine whether PG&E should be restructured completely. Should it be broken into separate gas and electric companies, or a number of even smaller companies? Should private investors continue to own it, or should ownership instead be held by the state or other public agencies with the wherewithal to manage it? Nothing was off the table, Picker had said at a commission meeting during which dozens of angry Californians urged him to do something—anything—to force PG&E into compliance. But Picker cautioned them that change would be slow, and difficult.

"One of the challenges of changing an institution like PG&E is that it has to continue to operate every day," he said. "This is really a little bit like remodeling an airplane in midflight."

The proceeding was an offshoot of an earlier one the commission had undertaken after San Bruno. It had spent months examining PG&E's "safety culture" and had commissioned a consultant to spell out its shortcomings. Safety culture is one of those nebulous terms reminiscent of signs reminding workers to hold the railing when descending the staircase, or to always wear the right protective gear. Most people fail to register what it really means. It's a vague-sounding concept, but a basic one. An organization with a strong safety culture keeps all aspects of its operations functioning without harming anyone by virtue of the fact that all of its employees value safety above all else. In April 2019, Picker convened a special meeting. It was structured as a workshop to address why PG&E couldn't achieve that, and what the commission could attempt to do about it.

The meeting opened with a presentation by Dr. David Hofmann, an expert on organizational behavior who had flown in from the University of North Carolina at Chapel Hill to explain how the culture within a company like PG&E could break down at various levels. He homed in on middle managers, employees like Bill Manegold and others who had taken the stand during the San Bruno trial. Hofmann set the stage with a simple example. The middle manager is responsible for overseeing two budgets: maintenance spending and capital spending. If the manager invests $1.00 in capital, he gets $1.20 back. Spending $1.00 on maintenance, meanwhile, is just $1.00 out the door—often with no obvious result or reward.

"People in my world call that a 'dynamic non-event,'" Hofmann said. "You have to spend a lot of money and do a lot of things for nothing to happen."

On top of that, Hofmann said, the manager is sitting in a comfortable office in San Francisco or Sacramento, many miles removed from the workers in the field, making it easier to become detached from the risk that results from shifting money around. Instead of spending $1.00 on maintenance, the manager decides to spend $0.80. Now there's $1.20 to invest in capital improvements, with the prospect of a larger return. The immediate consequence? None.

"Because I can easily lose sight of the harm, that decision gets easier and easier," Hofmann said. He called it a slow "drift into failure," a nod to the title of a book that explores how complex systems break down. In this case, it aptly described how an investor-owned utility like PG&E could put profits over safety for years with no consequence—until disaster disrupted the pattern.

Picker, his white mustache thick as a bristle brush, was the first to ask a question when Hofmann reached his last PowerPoint slide.

"So how do you interrupt this drift from outside?" he asked. "What are the effective tactics that reconnect all the pieces?"

"That's a really good question," Hofmann replied. "And I think it's really hard to do."

The hearing continued for two days, with a litany of other experts called to recommend how the CPUC might implement changes to the way PG&E governed its executives and managed its operations. But Picker wouldn't be the one to make a decision on what the agency ought to do. A few weeks after the hearing, he announced that he would step down sometime during the summer. He had known for months that Newsom wanted to replace him with an appointee of his own. Picker wasn't going to stand in the way. And heading the commission—with its plodding bureaucracy and ineffectual oversight—had frustrated him deeply. In assessing what he had been able to accomplish, he felt he had made some progress in expanding the commission's safety oversight, but not much else. He gave his performance a C across the board. Shortly after stepping down, he remarked that he'd had a worse job only once before, at a meatpacking operation. His task was to pick up dead cows.

Governor Newsom began the summer under enormous stress. Lawmakers had been scrambling for weeks to cobble together a solution to reduce the risk of more wildfires pushing PG&E, as well as Southern California Edison and San Diego Gas & Electric, to the brink of financial collapse. They came up with a multibillion-dollar fund that each utility could use to help cover future liability costs. The point was to assure Wall Street that investor-owned utilities had a future in California.

The legislation didn't directly address inverse condemnation, the provision holding the state's utilities liable for fires ignited by their equipment, but it created a pool of money that would essentially function as an insurance policy. A utility that had properly maintained its power lines would be allowed to take from the fund with no reimbursement requirement. But a utility that failed in that regard would be required to pay a deductible in order to tap it.

Money for half of the fund, valued at $21 billion, would be supplied by the utilities, each of which would contribute a set amount based on its size. The rest would come from their customers. There was already a funding mechanism in place: the surcharge tacked on to customer bills to pay off bonds that the Department of Water Resources had issued to purchase power on PG&E's behalf during its first bankruptcy. Rather than let it expire, lawmakers agreed to extend it to help mitigate the new electricity crisis.

For PG&E, there was a catch. Newsom stipulated that the company could access the fund only if it got court approval of its restructuring plan by June 30, 2020. It was a hugely challenging deadline. Less complicated bankruptcy cases often took years to resolve. The company's access to the fund depended on its ability to negotiate settlements with fire victims, as well as the pool of companies and hedge funds that held insurance claims. All of them would need to vote in support of a restructuring plan in order for PG&E to earn Montali's blessing.

PG&E had already reached one major settlement. In June, it had agreed to pay $1 billion to resolve claims filed by California municipalities and governmental agencies, including Butte County and the town of Paradise. The money would help them rebuild communities and compensate emergency response services for their work during the fires. It would be paid out of the relatively small pool of cash the company had on hand.

Meanwhile, the company had made little headway settling with fire victims, whose claims were still being filed. The process of determining their value was becoming difficult and controversial. The biggest sticking point was the Tubbs Fire. Trial lawyers insisted they had evidence proving that PG&E had started it. In mid-August, as Montali considered whether to permit a trial, he called for a hearing on how the claims should be estimated. It was a daylong affair. Scores of lawyers appeared to make their case for how the court should arrive at a number.

Chief among them was Kevin Orsini, who had earlier appeared to defend PG&E in front of Judge Alsup. He had also taken a lead role in the bankruptcy case, where he charged $1,500 an hour for his services. He told Montali that settlement discussions were well underway with insurers, one of the largest classes of claimants. But he warned of how difficult it would be to settle claims by individual victims. The company, in its discussions with the committee representing them, had offered $7.5 billion in compensation as an opening gambit. The committee scoffed. The victims would never settle for that amount. Any plan of reorganization would need to offer many billions more.

"Until we get Tubbs resolved, until we get into some estimation of the ultimate damages and guidance from a court, they're not even in a position to begin to evaluate whether to support a plan," Orsini told the judge.

In the back of the courtroom, one of the victims was listening. Will Abrams had been sitting there quietly, watching the proceedings and making note of how the attorneys addressed the judge. He desperately wanted the Tubbs Fire to go to trial. It was a crushing feeling, imagining that evidence of PG&E's culpability might go unseen and unheard. His family had been living in a small rental in Santa Rosa for nearly two years. A settlement would go a long way toward helping them rebuild.

It wasn't just about the money. Abrams hadn't been the same since the fire. He couldn't get through a day without thinking about his drive through the flames, hands tight on the wheel every time he got in the car. Some days brought vertigo, the sensation of spinning out of control until he vomited. His children were struggling, too. His young son felt it deeply. Every week brought another phone call from the school nurse as she tried to help him through another panic attack. In August, Abrams sent a letter to Montali, pleading for a Tubbs Fire trial. "Attention Judge Montali," it began. "My name

is William B. Abrams and I am a wildfire survivor who lost everything I own and ran with my family through the flames the night of
the Tubbs Fire . . . I urge you to please consider the financial impacts
on wildfire victims and survivors who want above all else justice and
a ruling that can be relied upon."

Abrams watched the letter appear on the docket a couple of weeks
later. For the first time in a long time, he felt a sense of agency. The
circumstances were so far beyond his control. But asserting himself
made them seem less hopeless. He began thinking about the attorneys who bickered in front of Montali each day. Many of them hadn't
spent a lot of time in Northern California, much less lived through
the fires that had forced hundreds of thousands of people from their
homes. On September 12, Abrams filed a second letter. He asked
Judge Montali for formal admission to the case. He wanted to testify
personally.

"Please consider that residents and wildfire survivors like me are
at a significant disadvantage without the resources of PG&E to make
their voices heard and influence this bankruptcy proceeding," he
wrote. "I think this wildfire survivor perspective would help all parties understand the tradeoffs and compromises more fully."

The very next day, PG&E announced its second major settlement. It agreed to resolve claims by insurance companies for $11 bil-
lion. The settlement negotiations were confidential, but some details
emerged in court filings. The insurance group had initially demanded
about $20 billion, a number PG&E disputed. So it agreed to a lower
amount under one condition: it wanted all cash, up front. Despite the
supposed compromise, many among the insurers still stood to make
hundreds of millions of dollars. The largest among them was Baupost Group, the hedge fund that had bought insurance claims at a
discount on the secondary market, some for as little as fifty cents on

the dollar. It was, at that point, on its way to amassing about $6 billion in claims, more than doubling the $2.5 billion it had purchased at the outset.

The two settlements with public agencies and insurers together drained the company of $12 billion in cash. What little the company had left was far less than what it owed to fire victims.

Eleven

BLACKOUT

The last thing Michael Picker did as president of the California Public Utilities Commission was to read a poem. As he ended his final meeting on August 15, 2019, he pulled out the second half of John Updike's "Telephone Poles," a single stanza on "the nature of our construction"—and what it displaces. He lingered on the last two lines:

> These giants are more constant than evergreens
> By being never green.

It could have been read satirically. But Picker read it in earnest, as an ode to infrastructure. "I say this to remind you of the value, not perhaps the beauty, but the strength and the structure that it brings that carries prosperity to California," Picker said. "We should all be proud of it."

The room echoed with applause.

Autumn brought an air of foreboding. The drought had ended, but trees were still dying, too long starved of water and drained of nutrients by bark beetles. Across the state, more than 163 million trees were dead. And the devil winds—the Diablos in the north, Santa Anas in the south—would soon begin to blow. Each of the state's large utilities had spent the spring and summer in feverish preparation, dispatching tens of thousands of workers to attack errant branches and check poles and towers for weaknesses. They had also been building networks of weather stations to track wind speeds and cameras to spot fires. For Southern California Edison and San Diego Gas & Electric, starting a fire meant risking their solvency. For PG&E, it meant risking survival.

Even in its precarious position, PG&E had fallen behind on much of the prevention work it told regulators it would complete that year. By mid-September, it had worked less than a third of the trees it had planned to trim, removed less than half of the dead ones with the potential to fall on its wires, and hadn't fully addressed many of the power line hazards it discovered during its inspection blitz. The company had succeeded in installing more cameras and weather stations. But it still had a lower density of them than its peers to the south, given its far larger service territory. As a result, its meteorologists had a less granular view of wind conditions as they worked to forecast fire danger.

On October 4, 2019, the meteorologists watched as pressure built over the Great Basin. The Diablo winds threatened to return with a vengeance. Within days, they would blow up and over the Sierras to bend trees, lift branches, and stress the hooks and plates holding up power lines. The company had never been more aware of how vulnerable its infrastructure had become. Whole swaths of Northern California were in danger.

The meteorologists consulted with electric operations managers. They were frightened. Weather models showed the prospect of an intense windstorm, the worst of the season. It threatened to be even stronger than the one that unleashed flaming cyclones during the fire siege that killed forty-four people almost exactly two years earlier. PG&E held an urgent conference call with emergency response agencies throughout Northern California. A consensus emerged. The risk of wildfire was untenably high. PG&E would have to cut power.

At dinnertime on October 6, PG&E activated emergency operations at its San Francisco headquarters, a cavern of screens tracking weather conditions across Northern California. The meteorologists forecast gusts between fifty-five and seventy miles per hour throughout the North Bay and the Sierra foothills. Even the areas immediately surrounding San Francisco were at risk, with winds expected at forty-five miles per hour or faster. That was the threshold at which distribution lines, the small wires serving homes and businesses, were at risk of failing. For transmission lines, the larger wires crossing long distances, the threshold was fifty-five miles per hour.

The meteorologists outlined the areas in danger. The footprint encompassed power lines serving hundreds of thousands of businesses and households. If the company shut them all off, more than two million people in thirty-four counties would be in the dark. No utility had ever deliberately attempted a blackout of that size. But the electric operations team struggled to see how they could limit the scope. The potential wind speeds were such that they would have to shut off not just distribution lines but also lower-voltage transmission lines, which meant that customers living outside the riskiest areas would also lose power because of the way the system was configured. PG&E's system did not have a high density of devices known as sectionalizers, which divide power lines into discrete sections that can be selectively shut off without affecting the others. And the condition of the system—the untrimmed trees, the unfixed hazards—left no room for error.

The warnings to customers began on October 7. Hundreds of thousands of them received vague messages warning of potential shutoffs. "Gusty winds and dry conditions, combined with a heightened fire risk, are forecasted in the next 36 to 48 hours and may impact electric service," they read. "If these conditions persist, PG&E may need to turn off power for safety. Please have your emergency plan ready. Outages could last for multiple days."

PG&E directed customers to its website to determine whether their home or business would lose power. Millions of people tried to pull it up to find an interactive tool that was supposed to check the status of a given address. The site crashed under the onslaught of traffic, loading as a useless blank page just when it was most needed. Phones in the company's call centers rang nonstop. PG&E didn't have enough time or people to answer all of them.

Meanwhile, employees were frantically trying to contact 30,000 customers who relied on electricity for medical reasons. The company maintained a registry of people who used respirators, electric wheelchairs, or dialysis machines, or who had conditions that required their homes to be kept at certain temperatures. Failure to reach them could endanger their health. The registry, like so many of PG&E's other records, was incomplete. But employees did their best. About 6,800 customers didn't respond to calls and emails, so PG&E sent people to knock on their doors. The company managed to alert most of its medical registrants of the impending blackout. Still, some 600 of them went without notice, either because PG&E lacked their contact information or because it didn't realize they resided within the scope of the shutoff.

The outages began just after midnight on October 9, on the two-year anniversary of the deadly fire siege. The lights first went out in the North Bay and the Sierras. The East and South Bay were next, and then parts of the Central Valley. Come evening, more than two million people were in the dark.

The best way to understand our collective dependence on electricity is to watch people go without it. Within hours, Northern California had come to a near standstill. San Francisco had been spared, but chaos erupted throughout the surrounding area as traffic lights and office buildings went dark. Hospitals in the East Bay rushed to move refrigerated medications to facilities with power. Nursing home residents tripped in the dark. Lines formed at gas stations, which were forced to ration supplies. Hardware stores were inundated as customers fought over generators and flashlights. The California Department of Transportation only just managed to install emergency generators to light the busy Caldecott Tunnel, a four-bore connector between Berkeley and Orinda that can't function without power. And PG&E's website was still down. It had come back up several times, only to crash again in a matter of minutes. The phones rang constantly, often to no avail.

The second day without power was even worse. More than four hundred schools closed their doors. Cell service faltered for those who had the means to charge their phones. Groceries spoiled. Restaurant workers purged food from walk-in freezers. There was hardly any ice to be found anywhere. Office buildings closed. Hotels canceled reservations. Doctors canceled appointments. Hospitals canceled surgeries. Economists estimated losses in the billions.

The world watched in shock as California, the fifth largest economy, sputtered so spectacularly, all because a utility company had chosen to turn off part of its system. That afternoon, as pundits made derisive jokes, Governor Gavin Newsom hastily assembled a press conference. Gray Davis's downfall had been rolling blackouts to ease the supply crunch during the electricity crisis. Newsom feared that his downfall might be this new brand of blackout, a last-ditch safety measure gone terribly awry. He tore into the company, voice tight with disdain.

"What's happened is unacceptable, and it has happened because of neglect," Newsom said. "It's happened because of decisions that were deferred, delayed, or not made by the largest investor-owned utility in the state of California and one of the largest in the nation. This current operation is unacceptable. The current conditions and circumstances are unacceptable."

And the blackouts weren't over yet.

⸺

Hours after the governor sounded off, a weary-looking Bill Johnson approached a podium set up at PG&E headquarters. He wore a black vest with fluorescent tape over his shirt and tie, a sign he had just come from the emergency operations center. A handwritten name tag peeked out of a clear plastic pocket. News cameras were rolling. At that point, power had been restored to only about a third of those who lost it. It was obvious, Johnson said, that the company had been wholly unprepared to manage such a shutoff of such scale. But he said he believed it had made the right call.

"We simply could not continue running parts of this system given the risk to public safety," he said. "We must have zero risk of a spark."

It took PG&E more than four days to end the blackouts. Many customers were without electricity for at least three. Before restoring power in any given location, the company sent workers into the field to check lines for damage. They snapped photos of branches lodged between wires, circuits grounded by fallen trees, and equipment that had broken in the wind.

The governor had appointed a woman named Marybel Batjer to replace Michael Picker as head of the CPUC. Batjer grew up in Carson City, Nevada, and started her government career in Washington, DC. She moved to California to help manage the state budget after the end of the Cold War crushed the aerospace industry, giving way to the recession that heightened the appeal of deregulating the utility industry. She toggled between California and Nevada, working in various legislative and casino industry roles, eventually becoming Arnold Schwarzenegger's cabinet secretary shortly after the energy crisis and later working under Jerry Brown. By the time Newsom took over, she was well versed in the particulars of California politics and adept at navigating the state's many departments and agencies. She hadn't focused much on utility regulation, but she was good at getting up to speed on complex issues. She became president of the CPUC in July 2019.

A week after the blackouts, Batjer convened an emergency hearing. All eyes were on her. For all her years in government, she had never adopted the understated look of a bureaucrat. She wore thick-rimmed glasses and big jewelry, her dark blond hair blow-dried into soft waves. She appeared at the hearing in a Louis Vuitton jacket and a long, four-strand pearl necklace with matching earrings that nearly grazed her shoulders. Her opening remarks were calm, but pointed. The blackouts were simply unacceptable in their scope, she said, and PG&E had failed its customers in executing them so poorly.

"I've only been at the CPUC a couple months, but it does not take long to see the privilege investor-owned utilities have, being a unique provider of essential services to the public," she said. "It has also not taken me long to see that some utilities fail to understand what a privilege that is."

Johnson sat before her, hunched over a table in front of the dais. His gray suit was just a few shades darker than his wavy hair, combed back and slicked with product. He was conciliatory, but firm in his defense of the company's decision.

"Let me assure you that we do not like to turn off the power. It runs contrary to the reason any of us ever got into this business," he said. "But as I look back at last week, one of the things that stands out in my mind is that we actually didn't have any catastrophic fires in Northern and central California."

Johnson acknowledged the company's failings. He promised it would bolster its website and its call centers and do more to tell customers when it needed to turn off the lights. It would install more sectionalizers, the devices that allow for a more surgical approach to shutoffs. It would cut more trees and insulate some of its wires to protect them from flying branches. In some communities at high risk, it would build self-contained power grids that could stay online even when the surrounding area went dark. The work was imperative, Johnson said, because more shutoffs were inevitable.

He asked a rhetorical question. How was it that PG&E found itself in this position, confronted with the morbid dilemma of starting wildfires or stemming the lifeblood of the modern world? It wasn't just about PG&E, he said. It was about the entire state of California, and indeed the entire West, where the risk of wildfire had grown considerably in a short span of time as a changing climate made conditions hotter, drier, and more flammable. In 2012, Johnson said, 15 percent of PG&E's service territory was at high risk of fire. That percentage had since more than tripled.

"We're working to reduce fire risk in a multitude of ways, but this will take some time," he said. "In the shorter term, the conditions are right for wildfire."

The commissioners took turns addressing Johnson. His remarks had begged a single question, and finally, one of the commissioners asked it. How long until PG&E could reduce, or even eliminate, the need for blackouts?

"I think this is probably a ten-year timeline to get to a point where it's really ratcheted down significantly," Johnson replied. "But at the same time we're doing this, the risk is not static. It's dynamic, and it goes up every year."

The winds returned in a matter of days. PG&E's customers, still reeling from the disaster a week earlier, would soon be in the dark again. The company's distribution lines were at highest risk of sparking, given their proximity to trees. PG&E could attempt to limit the scope of the shutoffs by carefully targeting individual circuits, each of which served pockets of different neighborhoods within cities and towns. But transmission lines were risky, too, and far more difficult to shut off without blacking out large numbers of people.

The company, after assessing the risks to the system, decided to cut power to several dozen distribution lines and a limited number of transmission lines, mostly lower-voltage ones. It also targeted a handful of the old 115-kilovolt lines carrying hydroelectric power through the Sierras. The company deliberately avoided cutting power to 230-kilovolt and 500-kilovolt lines, which together made up the backbone of the transmission system. Doing so would severely limit its ability to contain outages in the areas most affected by the winds. All told, the shutoffs would affect about 179,000 customers in the Sierra foothills, then the North Bay and the Bay Area.

On the night of October 23, as the company began switching off its lines, gusts topping out at seventy-six miles per hour began pummeling a 230-kilovolt line in Sonoma County. A small connector, similar to the hook that started the Camp Fire, broke, dropping a live wire onto the tall steel tower. Sparks showered the ground below, igniting what would become known as the Kincade Fire. The fire barreled out of control. It burned for thirteen days and destroyed nearly seventy-eight thousand acres. It was the largest fire in the history of Sonoma County, bigger than even the Tubbs Fire. This time, no one died. But homes and businesses were destroyed.

The line's failure was a bad omen. Crews had checked it during the inspection blitz and found nothing amiss. ("Sometimes things just break," Johnson would later say, rather bluntly, in response to questions about whether PG&E could have done more to prevent the fire.)

But whether the line was up to spec or not was, in some ways, beside the point. The winds had caused it to spark, raising difficult questions about the company's blackout strategy. Would it have to start cutting power to 230-kilovolt lines? They were, after all, among the oldest in the system, with more than half of them built between 1920 and 1950.

On October 25, the company restored power to all who had lost it. But the winds were building again. The company issued a notice: it might have to cut power to some two hundred transmission lines, including a dozen 230-kilovolt conduits all throughout the Sierras and the North Bay. The next day, the lights went out again, this time for more than three million people. For some, it was the fourth time losing power that month. And once again, it was a disaster.

It was a dark dilemma. PG&E could no longer provide safe and reliable service when the Diablo winds blew. That was supposed to be a both-and objective for every utility. For PG&E, it had become an either-or proposition.

PG&E's fear of causing more destruction was compounded by the effect another catastrophic fire would have on its ability to get out of bankruptcy. It could barely handle $30 billion in liability costs; any more could jeopardize its ability to function. Shutting off the power was an existential decision for both the company and its customers.

Newsom convened with his strike team, which he had assembled shortly after the Camp Fire to debate solutions to the PG&E problem. Even before the blackouts, the team had contemplated what it would take for the state to take ownership of PG&E. It was far from the preferred solution, but the blackouts had put it squarely on the table. Across the state, residents were hoping for a dramatic transformation of the company.

On November 1, after PG&E had finally restored power to all who had lost it, an angry Newsom once again addressed the people of California. He said he had ordered PG&E executives, as well as investors and lawyers active in the bankruptcy process, to appear in Sacramento the following week to discuss how to expedite the company's Chapter 11 exit by the June 30 deadline the state had imposed in establishing the wildfire fund. If PG&E's shareholders and bondholders couldn't agree on a plan of reorganization, Newsom threatened, the state would intervene.

"PG&E as we know it may or may not be able to figure this out, and if they cannot, we are not going to sit around and be passive," Newsom said. "The state will prepare itself as backup for a scenario where we do that job for them."

The governor was hardly the only one who had lost confidence in PG&E's ability to manage the system. All across Northern California, the leaders of the cities and towns that had gone for days without power were ruminating on whether their locales could somehow break free of the giant utility. The most prominent among them was San Francisco itself, the company's home since John Martin and Eugene de Sabla joined forces more than a century earlier. The city wanted to establish its own municipal utility, publicly owned and operated. The idea had been months in the making. Mayor London Breed and her staff had started exploring it as early as January, shortly after the company filed for bankruptcy. Within two months, the city had launched a formal study on whether going the municipal route would benefit residents, and what it might cost to do it.

The study cut to the heart of a larger question about the provision of electricity in the US, the majority of which is done through investor-owned utilities. If private capital is eliminated, then what? Public ownership, most often through a city or a town, is one of the only alternatives. There are hundreds of municipal utilities across the country, most of which are small. In many cases, their service territories are islands within those of the private utilities, which collectively serve about three quarters of the population.

The public ownership model emerged at the turn of the twentieth century as large, investor-backed companies like PG&E built out power plants and wires in pursuit of economies of scale. In some cases, public utilities formed in fast-growing cities wary of large corporations sweeping in to sell them power. In other cases, they formed to serve communities that had difficulty attracting private capital because they were simply too small and too remote to justify the expense of the infrastructure needed to serve them. But public ownership fell out of favor in many places as private utilities expanded their transmission networks, making it easier to deliver power to remote areas. Private utilities acquired dozens of municipal ones, folding them into their growing service territories and counting on their dense populations as sources of revenue to finance further expansion.

Because of that, private utilities have often fought to block cities and towns from defecting on the grounds that remaining customers would have to pay more in order to keep the system functioning. Residents of Sacramento, for example, voted in 1923 to purchase PG&E's local distribution lines and operate them through a municipal agency. PG&E fought the city for more than twenty years. The case went all the way to the Supreme Court of California, which, in 1946, at last determined that Sacramento had the right to proceed. PG&E was forced to sell.

San Francisco faced even steeper hurdles in its move to defect, a public vote of no confidence in one of its most iconic companies. It wasn't so much about the size of the loss—San Francisco, with a population of less than a million people, constitutes a fraction of PG&E's sixteen million customers. But the city had long been symbolically important to PG&E, home to major financial corporations and technology giants that feed power demand. Many of the company's transmission lines had been built to serve San Francisco and the surrounding area, and the company had spent huge amounts of money making sure they were among the most reliable within the system.

Come September, San Francisco completed its study and concluded that breaking free would be best for everyone. In bankruptcy court, it submitted a formal offer: $2.5 billion to acquire the network of power lines and substations serving it. Mayor Breed called the offer "competitive, fair and equitable," enough to help PG&E pay its dues while giving San Francisco a means to provide better service to residents. At the end of the month, Johnson met with Breed and other city officials to discuss the offer. He listened politely. But he wasn't interested.

On October 7, as PG&E notified its customers to prepare for sweeping blackouts, Johnson sent Breed a letter officially rejecting the city's offer, which he called "significantly below fair market value." The company, he wrote, disagreed that the city could better serve its residents in delivering electricity. On top of that, it believed that San Francisco's defection would "unnecessarily and unfairly" burden remaining customers with higher costs. And the company had other plans to raise the money it needed. The company's system, Johnson told them, was not for sale.

"Our financing strategy to emerge from bankruptcy does not envision selling off company assets," Johnson wrote. "We believe we can fairly resolve and fund all claims and other items through conventional financial markets."

Already, the company's shareholders and bondholders were each plotting their own strategies.

———

Dan Richard, PG&E's former head of government relations, watched the fallout with sadness. He had left PG&E in 2006, shortly after helping it through the energy crisis, amid frustration with Peter Darbee and the transformation process. But he had continued to monitor it from the sidelines while working on a new project: developing a high-speed rail connecting San Francisco and Los Angeles. The idea germinated in the 1990s and kicked off in 2008, when voters approved a legislative measure to fund construction. Governor Jerry Brown had appointed Richard to the board and tasked him with finding political and financial support for the project, which had become costly and controversial.

Richard, who'd had a front-row seat in PG&E's first bankruptcy, had started thinking about ways to help the company in 2018, after the fire siege. He still had the same warrior mentality as he'd had during the energy crisis, and PG&E had become his white whale. He had been discussing options with two attorneys, one of whom had led PG&E through its earlier reorganization, as well as an analyst versed in utility restructurings. They arrived at an idea: What if PG&E, instead of an investor-owned utility, became a customer-owned cooperative?

Public utilities are one alternative to private ownership. Nonprofit cooperatives are the other. Cooperatives were established after the Great Depression to serve regions that private utilities couldn't—or wouldn't—serve. Large swaths of farming and mountain communities, with residents loosely spread out over hundreds of square miles, didn't fit the emerging business model as investor-owned power companies built out their systems. So the Rural Electrification Act of 1936, a New Deal program, established cooperatives and provided them loans for the construction of electric distribution systems to serve those who wouldn't otherwise have access to power. There are still more than eight hundred electric cooperatives across the country, serving an average of eight customers per mile of distribution line, roughly a quarter of the number served by private utilities.

The idea to transform PG&E imagined a very different sort of cooperative, one with a huge service territory and a dense base of customers who would together own it. Its shareholders would be bought out through the bankruptcy process, their need for dividends eliminated. It would be free to set its own rates. To Richard and his team, the benefits were obvious. PG&E, at the time, anticipated having to raise tens of billions of dollars over a decade to make its system safer, plus more for standard upgrades and additions. As a cooperative, PG&E could take money that otherwise would have gone to shareholders and invest it in the system. And it would be able to raise money more cheaply, by issuing low-interest bonds. That would save billions of dollars in financing costs and reduce the need to raise rates.

That was critical, Richard thought, because the problems the company faced were only getting more acute. Climate change was exacerbating bouts of heat and drought, making the forests drier and more flammable. In a pitch explaining the benefits of a cooperative, Richard's team pointed out that a changing climate had "necessitated a rethinking of the utility model." Already, PG&E's rates were among the highest in the country, and the company would have to keep raising them in order to address the growing risks it faced. For that reason, they wrote, the company should reorient itself to solely benefit its customers rather than the shareholders who stood to make millions on every capital investment it made.

Richard had been talking for months with Sam Liccardo, the mayor of San Jose. They had met through high-speed rail circles and had an immediate rapport. Liccardo had been fed up with PG&E for months. Even before the blackouts, Liccardo had been deeply skeptical of its ability to execute such a strategy. The October disaster, which left some sixty thousand San Jose residents in the dark, shattered what little faith he had left in the company.

When San Francisco went public with its bid to buy PG&E's assets, Liccardo began talking about whether San Jose should try to break free, too. Richard scheduled a meeting with him a few days later to pitch the idea of a cooperative. Liccardo asked sharp questions. Richard followed up with more information. Within a week, Liccardo was on board. It was a huge coup. As mayor of San Jose, California's third largest city and home to Silicon Valley, Liccardo had a loud microphone.

On October 21, 2019, a few days after another round of blackouts, Richard and Liccardo broke the news of their plan in *The Wall Street Journal*. The shutoffs, Liccardo told the paper, had underscored the need to completely overhaul the utility. "I've seen better organized riots," he said. "This is a crisis begging for a better solution."

PG&E executives had been aware that Richard and his team had been shopping the idea of a cooperative. But the public announcement took them by surprise. The company responded with a terse statement, similar to its response to San Francisco's bid. "We have not seen the proposal," it said. "However, PG&E's facilities are not for sale."

Perhaps it didn't matter. At that point, PG&E didn't have the final word on its plan of reorganization. That fell to Newsom and Marybel Batjer, the head of the CPUC. Any plan would need their blessing before final approval by the court. Liccardo began calling up other mayors, all throughout the state, to rally broad support for transforming PG&E. By November, twenty were on board. The mayors of some of California's most prominent cities and towns—Sacramento, Oakland, Berkeley, Davis—all wanted to see PG&E become a cooperative. They sent a letter to Batjer, asking her to seriously consider determining whether PG&E's plan of restructuring should be approved. The cooperative model, they wrote, would "allow PG&E to begin the process of restoring public confidence, in part by allowing the public to have greater role in determining decisions that increasingly have come to define matters of life and death." So many of PG&E's customers had come to believe that the company had put profits over safety. The company would never succeed in regaining their trust.

"There is a better way, and we want you to consider it," the mayors wrote. "Your proceeding is that opportunity."

Newsom's strike team spent hours at the whiteboard, gaming out the options for PG&E. The question wasn't simply who should own

PG&E. The scope of the debate went well beyond that. Any effort to change PG&E's ownership model would have economic implications for all of California.

One of the biggest challenges to a public takeover was the sheer cost of securing ownership. California would have to find tens of billions of dollars to buy out PG&E's shareholders and acquire its assets. That would likely require establishing a new state power authority to issue bonds to finance the purchase, and the debt would fall to customers to repay over time. A cooperative would use debt financing to the same end.

Transforming PG&E into either a public utility or a cooperative would eliminate shareholders and their need for dividends, freeing it to reinvest its revenues in making the system safer. And both would have access to capital at lower costs, potentially reducing rates. But neither change in structure would exempt the utility from inverse condemnation. Its customers would become responsible for problems with the grid and liable for fire damage. The utility's power lines, no matter the ownership, would inevitably spark again. The risk would never be zero, and it would go up every year.

Perhaps the biggest hurdle of all was the one San Francisco encountered in its ill-fated attempt to buy the power lines serving the city. PG&E was not for sale. Any bid to purchase the whole system would face staunch resistance from shareholders, delay PG&E's exit from bankruptcy, hold up compensation for fire victims, and spook investors in the state's other large utilities. A takeover was possible, but not without a fight.

Ultimately, Newsom chose the status quo. He struck a deal with the company to support its bankruptcy plan—if it used shareholder funding to reduce its debt load, held off on paying dividends for a few years, and submitted to more stringent regulatory oversight that could, in extreme circumstances, put a state takeover back on the table. There was another catch: if PG&E couldn't obtain court approval of its bankruptcy plan by June 30, its key to accessing the wildfire fund, it would have to put itself up for sale in a process that would likely invite another investor-owned utility to acquire it. Wall Street ownership would endure.

Twelve

A FRAGILE DEAL

On October 8, 2019, the day before PG&E's first big blackout, Will Abrams typed up another letter to Judge Dennis Montali.

"Unfortunately, it seems that wildfire victims will commemorate this two-year anniversary of the PG&E North Bay wildfires with a prolonged power outage," it began. "As communities scramble to prepare because PG&E has not successfully mitigated wildfire risks . . . I want to provide some additional perspective."

It was a lengthy letter. At the heart of it, Abrams didn't understand how the judge could approve PG&E's plan of reorganization if it didn't hold the company accountable on a clear set of safety metrics. Otherwise, what was to stop it from causing another deadly fire?

"I urge Your Honor to consider the likelihood of PG&E ending up right back in bankruptcy," he wrote. He filed the letter to the docket.

The judge hadn't acknowledged the two letters Abrams had filed in September, including the one requesting permission to testify in the case. Abrams had figured out how to intervene at the California Public Utilities Commission, where he had been trying for months to make an argument for more stringent safety metrics. He wondered if he couldn't just show up to bankruptcy court to do the same. He had heard enough of the proceedings to get a feel for the process. He checked the docket. The next hearing was scheduled for October 23 at 10:00 a.m.

That morning dawned with the threat of another PG&E blackout. The court reporter prepared to take down the names of everyone who showed up to speak. Lawyers for PG&E, its shareholders, and its bondholders all appeared, as did a slew of other attorneys representing banks, insurers, hospitals, state agencies in California, and the United States of America. The reporter listed twenty-two of them, diligently making note of their law firms. Many had New York addresses. At the very end of the list, she made a note: "Also Present: Will Abrams, Claimant, Fire Survivor." The hearing began promptly.

"Good morning," Montali said. "Crowded house."

The proceedings, at that point, had reached a fever pitch. In September, PG&E had come up with a tentative plan of reorganization with the support of its largest shareholders. The plan proposed settling victims' claims for just $8.4 billion, far less than the many billions of dollars their attorneys believed they were owed. To raise all the money it needed to settle claims and restructure its balance sheet, the company proposed issuing some new equity but far more debt. That would protect its shareholders from taking a big hit on the value of their holdings, a highly unusual outcome in bankruptcy, where shareholders are usually wiped out. Two of the hedge funds that had handpicked its board of directors—Knighthead Capital Management and Abrams Capital Management—had thrown their weight behind the plan, as had Baupost Group, which had acquired a substantial stake in the company alongside the insurance claims for which it would be paid in cash.

Knighthead, Abrams, and Baupost were big names in the world of hedge funds, where investment managers attempt to outsmart each other in finding troubled companies, buying their equity or debt, and selling it later at a profit. It's a high-risk, high-reward game, and it's often a short-term play—the smartest players push to make fast changes to boost the value of their target investments. Each involved in PG&E had shrewd people at the helm. Knighthead cofounder Tom Wagner, a former Goldman Sachs trader, had spent his whole career in the distressed investment business and often appeared on Bloomberg and CNBC to talk strategy. Abrams was run almost single-handedly by a secretive billionaire named David Abrams, who had earned a reputation for quietly delivering outsize returns to his investors. Baupost CEO Seth Klarman was the most famous of all, having been dubbed the "Oracle of Boston" for developing a Warren Buffett–like investment approach that, with patience, yielded big rewards. Together, the funds managed billions of dollars, with Baupost the largest among them. Their stakes in PG&E hung in the balance of the bankruptcy plan.

PG&E's bondholders were pushing a competing plan that upped the ante, offering to settle victims' claims for $13.5 billion by forcing the company to issue a large amount of stock. That would wipe out the company's existing investors by diluting the value of its outstanding shares and allow the bondholders to seize control of the company by purchasing the new equity. PG&E, meanwhile, would emerge from bankruptcy with less debt than the shareholders were contemplating, potentially giving it more leeway to make investments in the safety of its system.

Leading the competition was Elliott Management, one of the company's largest bondholders, as well as other prominent firms, including Apollo Global Management, a giant in the world of private equity. Elliott, led by an investor named Paul Singer, was one of the most aggressive and ambitious hedge funds in the industry, notorious for staking out prominent positions in troubled companies and then loudly demanding change. Perhaps most famously, the fund bought Argentinian bonds after the country defaulted and then held one of its ships hostage in Ghana in 2012 while seeking payment on the debt.

In PG&E's bankruptcy case, tensions were mounting among its many investors. Briefs filed with the court were rife with finger-pointing, with almost every group of stakeholders accusing the others of attempting to unduly enrich themselves.

"Let me just make an opening comment that's related to nothing except everything," Montali said at the start of the October 23 hearing. "I noticed in the last round of briefing that there's a little bit of emotion that's creeping in the words and lines, so I'm going to ask counsel to downplay the energy a little bit and stop making references to hedge funds or opponents by their name when it isn't relevant; Elliott on the one hand or Baupost on the other and so on . . . I just want to keep our arguments a little bit more balanced and civilized."

The hearing had gone on for nearly two hours when Will Abrams spotted an opening. He was still worried that PG&E, in its haste to get out of bankruptcy before the state-imposed deadline of June 30, wasn't taking a holistic approach to restructuring, focusing on dollars and cents without accounting for the risks of future fires. He approached the podium.

"Good morning," Montali said. "Still morning," he added.

"Your Honor, my name is Will Abrams. Respectfully, a little concerned about the schedule . . . ," Abrams began.

"Just tell me who your client is," Montali interrupted.

"Sure," Abrams said. "I'm a party to the CPUC proceedings, a wildfire survivor, and a claimant as well, and just very concerned that the schedule associated with the proceedings isn't taking into consideration . . ."

"Sorry . . . what?" Montali asked, processing that Abrams was not, in fact, a lawyer.

"Part of what I don't see in any of the plans that are being proposed is risk mitigation and what's going to occur there," Abrams said.

"We're only talking about timing for briefing some very legal questions at the moment, Mr. Abrams," Montali replied. "I can't deal with other matters."

Abrams pressed on. "We're moving very quickly to meet an arbitrary deadline while at the same time missing the mark in terms of focusing on safety."

Montali sighed as Abrams continued. "Mr. Abrams, I think you're missing the point," he finally said. "I'm trying to schedule how to decide whether this plan can be confirmed. Safety is important; it's not necessarily something I can deal with."

Montali was limited in his jurisdiction. Bankruptcy court is a place for financial restructuring, not safety restructuring. That seemed ironic to Abrams. How could the court confirm a plan meant to stabilize a utility like PG&E without directly accounting for the biggest risks to its balance sheet?

"I understand, Your Honor," Abrams said. He retreated from the podium, where he was replaced by two lawyers representing four hedge funds and a mutual fund adviser.

It was then that Abrams realized that bankruptcy court was, at its core, a practice in rote procedure. He couldn't just show up as a concerned citizen, like he could at the California Public Utilities Commission. So he went home and started googling how to intervene. Online, he found templates for court filings and the definitions of all the types of entries that kept appearing in the docket. He would become a movant. (That, he learned, was a party that files a motion.)

In the bankruptcy proceeding, fire victims were represented by the Tort Claimants Committee, a group of eleven victims who had been vetted by the US trustee overseeing the case. They were appointed to champion the financial interests of roughly seventy thousand other individuals and businesses with claims against the company. The committee itself had its own attorneys, as did each of its members, all of whom would join their clients at the negotiating table.

At the end of September, PG&E hadn't made any headway in negotiations with the committee and its attorneys, who had come out strongly against the company's proposed plan of reorganization.

They were flatly unwilling to accept just $8.4 billion in compensation. And they were especially upset that the insurers' group had succeeded in negotiating an all-cash settlement, making it likely that victims would have to take shares in the company if they wanted anything more than what PG&E was offering. Insurers and hedge funds, the attorneys noted, were in the very business of capitalizing on risk, and yet their settlement was devoid of it, foisting the risk instead upon the class of claimants with little ability—and even less desire— to handle it.

On top of that, the committee felt as though it had been sidelined as PG&E reached the settlement with insurers and hammered out its plan of reorganization. During a late September hearing, an attorney told Judge Montali that it wasn't that the committee had reached a stalemate with PG&E. It was that the company hadn't even attempted discussions.

"I'd like to tell Your Honor how many drafts of the debtors' plan the TCC saw before it was filed; the answer would be zero. How many meetings the TCC had with the debtors to discuss the plan; the answer would be zero. It would also be zero if Your Honor asked me how many meetings the debtors had with the TCC in the entirety of the case. That's how they're playing it," she said. "This is not a chess game."

It was quickly turning into one. The committee had, at that point, allied with the bondholders, the group of lenders led by Elliott Management, because their plan offered more money: $13.5 billion. It was a big blow to PG&E and its shareholders, which needed the support of fire victims in order to get court approval of their plan. So PG&E and its shareholders, in frustration, did something that would change the course of the entire proceeding. They tapped a man named Mikal Watts to end the standoff.

In the billboard-covered world of personal injury law, Watts was a prominent figure. He was a bald, heavyset man with a distinct Texas accent, a Corpus Christi native who graduated from the University of Texas School of Law at just twenty-one years old. His San Antonio law firm, Watts Guerra, had gotten involved in a number of high-profile disasters, notably the 2010 Deepwater Horizon explosion. He and his team negotiated a $2 billion settlement with BP to compensate tens of thousands of Gulf Coast fishers whose livelihoods had been affected by the massive oil spill. The settlement turned controversial in 2015, when Watts was indicted on federal charges of conspiring to falsify thousands of those claims. He was ultimately acquitted of the charges the following year in a trial in which he represented himself.

Watts homed in on PG&E just days after the 2017 fires. His firm mounted an aggressive advertising campaign in the North Bay, promising maximum compensation to those who had lost family and property. He did the same thing in the town of Paradise after the Camp Fire, organizing dozens of town hall meetings where he delivered folksy, congenial pitches for his firm. Erin Brockovich spoke at several of them. Hundreds of people showed up.

The campaigns were hugely successful. Watts wound up representing some sixteen thousand fire victims, more than any other attorney. And he stood to make the most money of any of them. He charged a 33 percent contingency fee. If he succeeded in securing a settlement for his clients, he would keep a third of the money. (If he didn't succeed, he wouldn't get paid.) A multibillion-dollar settlement would net hundreds of millions of dollars for his firm.

Watts had scarcely been involved in the settlement process—none of his clients were on the victims committee. But PG&E and its shareholders saw Watts, with his huge client roster and bulldog tactics, as the person who could turn the negotiations in their favor. PG&E had, by the end of September, engaged the committee on the terms of a settlement, but progress had been slow. Attorneys conferred privately with Watts and asked him to step into the fray. He began calling up other trial lawyers who collectively represented thousands of fire victims.

In October, Watts helped to establish a group of thirteen law firms to intervene in the committee's negotiations. Together, the firms represented more than two thirds of the seventy thousand victims suing the company. To the committee, the larger group seemed to have appeared out of nowhere. There were suddenly new faces at the negotiating table. An attorney representing one of the committee members was stunned the morning they first appeared during a closed-door mediation session. "I said, 'Who the hell are these guys?'" he later recalled. "It was the first time I ever heard of them."

One of the victims on the committee was Kirk Trostle, who had been anguished to learn about the Camp Fire during his time in Hawaii and returned to California to find his entire town destroyed. Trostle, a longtime Butte County resident who had retired after many years with the Oroville and Chico Police Departments, had been proud to join the committee. The Camp Fire had displaced almost his entire family, as well as dozens of friends. Representing their financial interests in such a complex case seemed to him like a noble thing to do.

Trostle had taken his job seriously. He spent months studying financial and legal jargon so that he could participate in negotiations. When the larger group of lawyers swept in to steer the conversations, he began to feel his efforts had been futile—and realized how little power the committee actually had.

The newcomers, for their part, felt they were better positioned to argue on behalf of victims by virtue of the fact that they represented the majority of them. They commandeered negotiations, with Watts taking the lead. Several meetings in late November produced a win for the shareholders: a rough outline of a settlement that essentially blocked the bondholders' competing plan. On December 6, a Friday, the deal was done. PG&E would compensate victims by establishing a $13.5 billion trust. It would fund half of it with cash. The rest would come in the form of shares in the company.

Two days later, Watts was feeling pretty pleased with himself. He was at the Flamingo Resort, a three-star hotel in Santa Rosa just a short walk from IHOP and Safeway. Watts, a natural showman, was about to give a presentation to a couple hundred of his North Bay clients. They had assembled in a curtained conference hall to hear about the deal he had just struck with PG&E and its shareholders, pushing the bondholders aside. Watts had done more than any other attorney to shepherd it across the finish line. And he was going to tell his clients exactly how he did it.

Watts ascended a small stage to stand beside a projector screen. He wore jeans and a simple blue button-up, glasses perched atop his forehead. He looked out over the audience, seated in rows of red upholstered banquet chairs. Many wore flannel shirts and ball caps. One woman brought knitting needles and casted on and off as Watts began speaking. He struck just the right tone.

"My wife asked me, 'Why are you goin' back out there again on a

Sunday?' And I said, 'Well, baby, these people are hardworkin' people, they've got Monday-through-Friday jobs. This is the only time I can talk to 'em,'" Watts said. "'And oh, by the way, it's after the 49ers game.'"

The audience laughed appreciatively. "Yeah!" someone exclaimed.

Then he explained why he was there. His job had been to negotiate on their behalf. Now, it was their job to decide whether they wanted to support the settlement. He would walk them through not only the terms of the deal but also how it came together, all in pursuit of what he called "informed consent."

"I work for you; you're my boss," he said. "You have decisions to make, and I think that they're obvious and that you want to make 'em, but that's not my call. That's your call."

He flipped through a set of strangely detailed slides. When he mentioned a settlement meeting, he included screenshots of calendar entries, or email exchanges with other people who attended. Things got serious during a meeting on November 18, Watts said. From there, it was a race to hammer out the details. Right after Thanksgiving, Watts was in constant communication with Knighthead cofounder Tom Wagner, who held perhaps the most influence of any of PG&E's shareholders.

"To be blunt, I never met the guy before six weeks ago. But I'm here to tell you that if he's one of the new owners, PG&E is in good hands," Watts said. "He's really smart and I trust him."

Constant communication among all parties is normal during negotiations. But Watts and Wagner appeared to have an especially friendly relationship. Watts flashed a screenshot of an email he sent to Wagner with the subject line "WRESTLING WITH THEM, BUT MAKING PROGRESS," in which he had written that he was pushing lawyers representing the victims committee to "get off their asses" and get the shareholders' lead attorney an outline of the deal. Days of back-and-forth concluded with the settlement, formally called the restructuring support agreement, or RSA.

The deal had several significant stipulations. It included a confidential settlement for Tubbs Fire victims like Abrams, and dozens of other Santa Rosa residents who had shown up to the Flamingo Resort. That meant the issue of whether PG&E started the fire would never go to trial. The deal also valued victims' compensation at $13.5 billion, allowing PG&E to forgo a lengthy and likely frustrating estimation process to determine exactly how much it owed them. Watts told his clients that he believed the total was much higher. But the company had very little to give at that point. Demanding too much, Watts said, could cause a Chapter 11 reorganization to become a Chapter 7 liquidation. In that case, everyone would risk going unpaid.

Then he turned to what would soon become the settlement's biggest sticking point: PG&E's plan to fund the trust. "It was not lost on me that I was going to have to sit in the Flamingo Hotel without a flak jacket and explain to two hundred people why it is that I agree we should take PG&E stock," he said. "I mean, they burned down your town. Why would that be okay?"

His answer was simple. Watts said that unless the trust was funded in part with shares in PG&E, victims would receive far less in overall compensation. The company had agreed to put $5.4 billion in cash into the trust upon emerging from bankruptcy. It would contribute another $1.35 billion in cash using certain tax credits available to companies whose expenses exceed their revenues. The remaining $6.75 billion would come in the form of equity. The trust would hold nearly 21 percent of the company's outstanding shares at the outset.

"You can only squeeze so much juice out of a turnip," Watts said.

Then Watts shifted to a part of his presentation that constituted an important legal disclosure, the significance of which was likely lost on most everyone in the audience. He pulled up a slide titled "What to Watch Out For—Protecting Yourselves." With the click of a remote, a subtitle appeared: "Litigation Funding—Conflicts of Interest."

"I want to disclose something to you," Watts said. "There are all sorts of conflicts of interest out there that in assessing what I tell you, you need to consider." He would know, he said, because something strange had just happened to him.

It started in September, when Watts said he got a call from some lenders with Stifel Financial, a Missouri-based investment bank. They had heard he was representing PG&E fire victims, so they offered Watts a $100 million loan. Lawyers like Watts who work on contingency don't get paid until resolving their cases, so many borrow money in the interim to help fund their litigation.

"Next thing you know, I've got this huge line of credit," Watts said. "I don't really know who they are, but they're nice."

Litigation funding is a touchy subject in the legal realm. It often raises ethical questions, and the rules are murky. There are several types of arrangements. In some cases, investors agree to back litigation in exchange for a percentage of the attorney's fees and, often, the ability to influence his or her litigation decisions, creating a serious risk of a conflict of interest. In other cases, the financing is offered on terms similar to those of a credit card—the attorney has to pay back the money with interest. Watts professed to have the latter arrangement.

The Stifel lenders asked Watts for a meeting. He flipped to a slide showing a photo of the bar at the Park Hyatt, a five-star New York hotel where he had met them for drinks. There, they introduced Watts to two investors with Centerbridge Partners, a private equity firm that held shares in PG&E.

"Real nice guys," Watts said.

The next day, Watts got a call from a Centerbridge portfolio manager, asking for a meeting. Watts flashed a photo of a man in a shirt and tie with dark eyebrows and very close-cropped hair. He pulled up a chatty email in which he had offered to change his flight to San Francisco to accommodate the man's schedule.

Within days, Watts said, he realized that Centerbridge had become one of his lenders by purchasing part of his debt through the line of credit. So had Apollo Global Management, one of the bondholders that had allied with Elliott Management in its effort to trounce the shareholders with a competing plan to get PG&E out of bankruptcy.

"I said, 'Holy moly, okay, I know what's goin' on here,'" Watts said. "These guys are trying to play me."

He described the game as a battle between equity and debt, with Apollo and Centerbridge each working to push the plans put forth by their respective teams. It was the Centerbridge portfolio manager who introduced Watts to Wagner, the Knighthead cofounder who led negotiations on behalf of PG&E's shareholders. The two had an instant rapport and began talking regularly, giving Watts greater insight into the equity plan. Then a portfolio manager at Apollo introduced Watts to Jeff Rosenbaum, Elliott's senior portfolio manager orchestrating the bondholders' plan. Watts got along well with Rosenbaum, too, he told the audience, at one point hosting him at his box at a University of Texas football game. Watts took over negotiations with the bondholders, too, and said he pushed to see what they had to offer victims.

The shareholders put forth their offer in a matter of weeks. Even more so than the bondholders, the shareholders had a key objective: to cut a deal with victims quickly. Delaying the process could jeopardize the company's ability to get court approval of its plan by June 30. In that case, PG&E would be barred from participating in the wildfire fund, exposing it to huge potential liability costs that could wipe out the value of its equity. Even the simple act of missing the deadline would likely tank PG&E's share price, making it harder for investors to sell their shares at a profit.

Just hours before Watts and his team signed off on the shareholder deal, the bondholders improved their offer: $13.5 billion, with $7.75 billion in cash, roughly a billion more than the shareholders could offer. It would later sweeten the deal even more, by offering to pay the entire amount in cash by issuing more equity at the expense of shareholders. But Watts wasn't persuaded. The deal involved a number of stipulations that Watts believed would delay the resolution of the case by deadline, a risk he considered unacceptable. (The bondholders, for their part, said they planned to meet the deadline, too.) On top of that, Watts added, he didn't trust Elliott Management and its founder, Paul Singer, whom he called a "destroyer."

"I don't want to get into details about why I feel this way, but all I will tell you is to google 'Elliott Capital Management and Paul Singer,'" Watts told the room, knowing they would find plenty of articles that portrayed Singer as a predatory investor. It came across as a scare tactic to bolster the message: the bondholder plan, despite the extra cash, was off the table.

"Google what again?" a woman shouted.

On December 16, about a week after the attorneys signed the deal, Will Abrams put the final touches on a court filing. It wasn't a colloquial letter to Judge Montali. It was a formal objection to the settlement agreement, formatted just like all the other objections that had appeared in the docket.

"William B. Abrams ('Claimant') hereby objects to Debtors' Motion Pursuant to 11 U.S.C. §§ 363 (b) and 105 (a)," it started. "As set forth in more detail below, the RSA is unreasonable for the Debtor to propose and not a prudent exercise of business judgment."

The objection had three main points. Abrams believed there was no way victims would be made whole if they accepted the settlement and all the risks it carried. He couldn't stand the idea of forgoing a Tubbs Fire trial. And he still believed that any deal was fatally flawed if it didn't address how PG&E should handle the wildfire and climate risks that had driven it into bankruptcy in the first place.

The next day, Abrams woke up early, knowing it would take more than an hour to get from his rental house in Santa Rosa to bankruptcy court in San Francisco, where Montali was set to decide on whether to approve the settlement. Still, he wound up rushing out the door. He had put on a suit, but in his haste, had forgotten to grab his dress shoes. He tiptoed into Montali's stuffy courtroom in his black Birkenstocks and took a seat next to a lawyer for PG&E's bondholders. The lawyer, wearing slip-on loafers, looked askance at Abrams's feet. "And I thought I wore comfortable shoes," he said.

Abrams waited patiently as lawyers for PG&E, the victims committee, the insurers, the bondholders, and the shareholders debated highly specific aspects of the deal. About halfway through, he stood up.

"The gentleman who was standing up, were you here for the City of San Jose?" Montali asked. Abrams shook his head and reintroduced himself.

"Oh, Mr. Abrams. Okay. Mr. Abrams, yeah, go ahead if you want to make a brief statement," Montali said. But the judge didn't have much patience. "A lot of your arguments really aren't something that I can deal with today."

Abrams stood to make his case. More so than ever, he was concerned about the speed at which everyone was moving, all in service of securing PG&E's access to the wildfire fund. The victims' settlement had been hammered out in a matter of weeks. With more time, Abrams thought, it could be improved. And he wanted the chance to assess all the options. The bondholders, after all, had offered more cash.

"What I heard being said is that the opposing plan comes forward with a better package for wildfire survivors, but they will never see it," he said. "The attorneys from the TCC have to represent that this is the best thing since sliced bread. And that will give the wildfire survivors very little information to make a very complicated decision."

"Okay," Montali replied, after a bit more back-and-forth. "Mr. Abrams, thank you very much for your comments."

After several more hours of discussion, Montali reached his decision.

"I don't think that I have the wisdom or the knowledge . . . to second-guess the decisions of those victims who have told their lawyers, this is how we want to go with the plan," he said.

He rubber-stamped the $13.5 billion settlement. Victims would next have to vote to approve it.

The anguished letters began appearing in January. Suddenly, it wasn't just Abrams who had taken an interest in the bankruptcy court proceedings. Other victims, too, began writing to Montali, urging him to renege on his earlier decision and reject the settlement. Many of the letters included photos of people living in tents in Paradise, unable to find or afford temporary housing as they waited on compensation from PG&E. They all agreed: $13.5 billion wasn't nearly enough if half of it depended on the company's share price.

"There are so many things about this current proposal that are unacceptable: Insurance companies getting top priority and $11 billion, even though they are insured themselves against such disasters and paid by premiums to cover fires," one of the letters read. "Stock options that may be worth NOTHING in the near future."

The construct of the settlement had precedent. Other cash-strapped companies facing massive liability costs had paid claims through trusts funded in part with stock and other assets. Asbestos manufacturers, for example, funded settlement trusts to compensate workers who developed cancer and other health problems, as well as those who would fall ill in the future. But the trust PG&E had proposed carried considerably greater risk. The companies facing asbestos claims had stopped manufacturing those products, limiting their liability exposure going forward. PG&E's power lines, meanwhile, would continue to ignite more fires, jeopardizing the future value of the stock in the trust.

For weeks, Abrams had been fixating on what Montali said at the end of the December hearing, when he pointed to victims' support of the settlement as the reason he ought to approve it. Abrams was sure that most of them had no idea what was going on as the attorneys hashed it out. On New Year's Eve, he created a SurveyMonkey poll and asked a few victims' support groups to distribute it to their members.

"Were you told by attorneys prior to the December 17th ruling that the RSA had a number of compromises including acceptance of partial payment in PG&E stock, eliminating Tubbs Wildfire Court Trial and locking out any other existing or new plans that might offer claimants and the public a more favorable settlement?" one of the questions asked.

Forty-six respondents answered yes. One hundred and ninety-two answered no.

At the end of January, Abrams filed a lengthy objection that included his SurveyMonkey results. The bondholders, at that point, had dropped their opposition to the shareholders' plan in exchange for more favorable treatment of their bonds, clearing the way for PG&E to finalize its overall plan of reorganization. Within days, Montali would hold a hearing on whether to approve it. Abrams, in his objection, urged him to deny it, arguing that victims had unwittingly been roped into a deal that tied their compensation to the performance of a fundamentally untrustworthy company.

At the hearing, Montali invited Abrams to speak first. But the judge hadn't found his objection persuasive.

"I read again last night your objection to today's motion, and you have a number of points that you, as a citizen and as a fire survivor and as a customer of PG&E, have every right to complain about,"

Montali said. "You raise, in no order, the following uncomplimentary terms: 'mismanagement.'"

"Yes," Abrams said.

"'Political lobbying,'" Montali continued.

"Yeah."

"'Criminal behavior.'"

"Yep."

"Et cetera, et cetera, 'take the money and run,'" Montali said. "My point is that all of those are provisions and positions that you are welcome to take. But I didn't see anything specific that keys those arguments to today's agenda. So I don't want you to waste time telling me your views about hedge funds or the management of PG&E."

Abrams had anticipated Montali's skepticism. He told the judge he had been googling rules associated with approving restructuring support agreements. The judge was amused.

"Did you find any?" Montali asked. "There aren't a whole lot of them, you know."

In fact, Abrams had. One suggested that a court should approve an RSA only if the debtor had "exercised reasonable business judgment" in crafting it.

"There's nobody who could reasonably come to this podium and say PG&E's a company where you should take their word that they're following good business judgment," Abrams said. "They are the poster child of a company that should not be afforded that path."

Montali wasn't convinced of the need to reverse course. He reminded Abrams that every victim would have the chance to weigh in on the settlement. In order to reject it, a third of an estimated seventy thousand claimants would have to vote against it. In that case, PG&E would either have to go back to the drawing board or

persuade Montali to approve the plan anyway, something he would
have been disinclined to do. "It's not the voting day for you and thousands of other people," Montali said. "And if they vote the plan
down, there are consequences."

By mid-March, circumstances had changed in a way few could have
foreseen. The coronavirus pandemic had closed the courthouse and
upended the stock market. Shares in PG&E fell from about seventeen to seven dollars and plateaued from there. Its value had rarely
been lower.

By then, PG&E had come up with a $59 billion plan to finance its
exit from bankruptcy. It would emerge with $38 billion in debt, up
from $22 billion when it first appeared in front of Montali, a remark-
able use of the restructuring process. Most companies emerge from
bankruptcy with lighter debt loads. The company also planned to
raise nearly $16 billion in equity. A group of hedge funds, including
Knighthead and Abrams, agreed to purchase about $9 billion of it
at substantial discount if the company failed to sell that much by
conventional means. The remainder would be issued to the victims'
trust.

At the end of the month, the committee tasked with representing victims in bankruptcy court lost one of its eleven members. Kirk
Trostle, the retired police officer whose home was destroyed in the
Camp Fire, simply couldn't support the settlement. But he believed
his role on the committee barred him from speaking out against it.
So he resigned. He had started to wonder if the members of the committee had just been pawns in the bankruptcy chess game, appointed
to make victims feel as though they had strong representation when

Wall Street held the strings all along. Two other members of the committee soon followed suit. Together, they wrote a letter to their fellow victims, urging them to vote against the plan.

"PG&E's plan is deeply defective," they wrote. "Unlike the $11 billion in cash going to insurance companies, hedge funds, and $1 billion in cash to some public entities, the plan has victims taking substantial risk."

Lawyers for the committee had been working for weeks to renegotiate certain details of the settlement to ensure the trust received stock valued at exactly $6.75 billion. As written, it risked receiving less, given the market downturn. They were working under enormous time pressure. Victims had until May 15 to vote on the plan. It would take several days to tally the results. A confirmation hearing was scheduled for May 27. The schedule was critical if PG&E hoped to get approval of its reorganization plan by the end of June.

Frankly, Trostle didn't care about the June deadline. At its core, the wildfire fund was a safety net for PG&E and its investors. Though it had been designed to benefit everyone by quickly lifting PG&E out of bankruptcy, it had ultimately done little for fire victims, other than provide support for a restructuring plan that tied their compensation to the company's share price. Trostle hoped that by rejecting the plan, victims would signal to the company that it needed to change the terms, no matter how long it took.

Trostle wasn't alone in his frustration. A number of attorneys had started to question whether the deal was tenable, given the state of the stock market and the very real possibility that the assets in the trust would ultimately be worth far less than $13.5 billion. One prominent law firm that had once supported the deal decided that it could no longer do so. It advised its clients—several thousand victims—to shoot it down.

"Unfortunately, the hedge funds that have hijacked these bank-ruptcy proceedings (all to their significant financial benefit), did not care about your concerns BEFORE the coronavirus pandemic, nor do they care about them now," the firm wrote in a letter to clients. "Your concerns MUST be kept in the forefront of this case."

Watts, meanwhile, was leading a relentless campaign pushing his sixteen thousand clients—and anyone else within earshot—to vote in favor of the settlement. He set up a website called Fire Settlement Facts and plastered it with videos of himself answering questions about the deal. One titled "What Happens If Fire Victims Vote This Plan Down" took aim right at Trostle, who had taken to social media to explain why he and the other two victims had resigned from the committee.

"If Kirk wants you to believe that in seven days, PG&E has time to improve the deal and try to win fire victim approval prior to the court confirmation hearing on May 27, I've got some oceanfront prop-erty in Arizona I'm ready to sell you. What's going to happen is the equity will dump the stock and cut their losses," Watts said in the video. "The funding will be gone."

For weeks, Watts and other attorneys who had allied with him had been hosting virtual town hall meetings in which they took questions and explained the terms of the deal. The takeaway was always the same: If the settlement falls apart, there is no plan B. No one gets paid.

In several of the meetings, Watts revisited what he had told cli-ents back in December, at the Flamingo Resort. He explained what had happened with Apollo and Centerbridge, the two firms that had purchased part of his line of credit from Stifel. A number of other attorneys were immediately suspicious. What Watts was describing sounded like a conflict of interest. At the very least, it appeared the two firms had bought influence in the case. If that influence amounted to a conflict, Watts would be subject to strict disclosure rules that would require him to get written consent from all of his clients acknowledging that they understood the circumstances. Mentioning it at a few town halls wouldn't be enough.

Watts came out vigorously in his own defense, arguing that his line of credit, despite the third-party investments, did not amount to a conflict because Apollo and Centerbridge had no influence over his decisions and couldn't share in his fees. He said he used it to fund all of his firm's litigation, not just the PG&E case, adding that he took care not to negotiate directly with Apollo and Centerbridge once he learned of their involvement in the loan.

Some attorneys questioned whether the line of credit came with terms that might have influenced Watts's desire to reach a settlement and resolve the bankruptcy as quickly as possible. People familiar with the deal pegged the interest rate at 18 percent. That could explain why Watts sided with the shareholders instead of the bondholders, whose plan offered more cash but might have taken more time to execute. Watts insisted the interest rate wasn't that high. Still, the longer the case dragged on, the more Watts would owe. One lawyer described it as "the silent elephant in the room." Others doubted the interest rate mattered, given that Watts had other revenue sources.

All of this was impossible to verify. Watts never disclosed the exact terms of the financing. Centerbridge never revealed how much it had purchased of the line of credit, while Apollo—which had been among the bondholders—said it owned less than a fifth of it and had no control over litigation decisions. On May 12, three days before the voting deadline, a group of fire victims and their lawyers mounted an eleventh-hour challenge in court. They asked Montali to consider whether the votes of Watts's clients should be discounted or recast, given that he had never gotten their written consent acknowledging the potential conflict. But time was running out. When asked to decide, Montali punted.

On May 22, PG&E announced the results of the vote. More than 85 percent of victims had voted in favor of the settlement. Almost none of Watts's clients had rejected it. The deadline for court approval of the restructuring plan was just over a month away. Montali would be the one to make the final call.

Thirteen

TWICE CONVICTED

few days after the Camp Fire stopped burning in November 2018, Butte County Deputy District Attorney Marc Noel put in a call to the California attorney general's office, looking for help with his investigation. A lawyer there gave him a tip: Call James Haggarty, who had led the federal investigation into the San Bruno explosion years earlier. If anyone knew where to start, he did.

In early December, Noel reached Haggarty by phone to ask if he would be willing to offer some guidance. Noel was quick to give him an out—the San Bruno case had been the most arduous in his career. If he never wanted to see another PG&E inspection record again, who could blame him? But Haggarty didn't think twice. He offered any assistance he could provide. Just as he had during San Bruno, he quickly became one of the prosecutors' greatest assets.

Haggarty advised Noel on the first step: Ask for every inspection record associated with the Caribou-Palermo, the transmission line that failed early on the morning of November 8. PG&E turned them over in December. On a cold January night, Noel was working late from home. His dining room had become his war room, the table stacked with notes and documents. At that point, Noel had read enough records to get a sense of what PG&E was supposed to do in monitoring its transmission system. A single line could have dozens of towers, so inspections could take hours or even days if done properly.

Noel squinted at his screen. The record he had pulled up documented what was supposed to have been a detailed inspection of the Caribou-Palermo, which had more than two hundred towers running fifty-six miles down the mountain to a substation south of Oroville. The inspectors recorded looking at all of them in what seemed to be an impossibly short period of time. Noel put out an email to the team, asking them to pull up the document. "Am I reading this correctly?" he wrote. Ten minutes later, Haggarty called Noel. "You're absolutely right," he said. The record seemed to indicate that inspectors weren't doing thorough work in assessing the company's transmission towers. It was one of the first times Haggarty saw a clear parallel to San Bruno.

Piecing together more evidence took months. District Attorney Mike Ramsey assigned about seven people to work on the case, and the California attorney general's office lent several of its prosecutors to the effort. Come spring, the team decided it was time to empanel a grand jury to issue subpoenas, examine witnesses, and determine whether PG&E should face criminal charges. Ramsey took the lead in assembling it. His challenge was to find jurors who could render an impartial verdict. Most everyone in the vicinity had been affected by the Camp Fire. Of one hundred candidates, he chose nineteen.

The grand jury started with a basic question: Did the Caribou-Palermo ignite the Camp Fire? The jurors reviewed much of the evidence that Cal Fire had used to arrive at its conclusion to establish the factual basis for their probe. Meanwhile, Ramsey and Noel were getting deeper into the complexities of PG&E's transmission system, amassing evidence that would shed light on the far murkier issue of why the company had allowed the line to fail. They were beginning to see more parallels with San Bruno. It appeared that in both cases, the company's inspection practices had failed to identify equipment at imminent risk of failure. The pipeline explosion had occurred after years of budget cuts that forced gas transmission engineers to resort to inspection methods that were cheap, and ineffective at assessing seam weld integrity. The question was whether similar financial pressures had affected inspections of the company's electric transmission system.

Noel and Ramsey called on a number of former PG&E employees for interviews. Among them was Bill Manegold, the former gas pipeline engineer who had testified during the San Bruno trial. Noel spent hours talking with Manegold about his testimony, homing in on how high-level pressure to cut costs had forced the gas transmission division to resort to cheaper inspection methods and run parts of the system at high pressure without having done the proper safety tests. Noel wrapped up his interview more certain than ever that a nearly identical dynamic had manifested within the electric transmission division.

It took Ramsey and Noel months to sort through the mess of documents that PG&E had turned over. They spent long days staring at their screens, working late into the night to try to make sense of what the company had delivered. Their requests had netted reams of engineering records, maintenance logs, policy handbooks, and regulatory filings. The company had also produced a cache of internal emails. Just as they had during the San Bruno trial, the conversations among engineers would seal the company's fate.

One day, Ramsey came across an email that stopped him cold. It was from an engineer asked to rank the risks associated with several transmission line projects. The company's ranking system, used internally to prioritize projects, had three categories: safety, reliability, and environmental, each with a maximum score of ten thousand. The total, out of thirty thousand, was supposed to indicate the risk the company faced in not completing any given project.

The Caribou-Palermo was among the projects the engineer ranked. He couldn't attend a meeting to discuss it, so he emailed his supervisor with his assessment. The Caribou-Palermo, the engineer wrote, should score less than two hundred because "there is no likely large environmental event (if structures fail, it will be likely due to heavy rain and no wildfires are possible then)." Reliability wasn't a concern, either, he said, because "although the likelihood of failed structures happening is high," such failure would likely cause an outage for less than a thousand customers.

"Also no likely public safety issue with live wires down because it is in a remote area," he wrote. It seemed he overlooked the presence of nearby towns, and the risk of fire spreading there.

Ramsey's heart sank. "Oh," he thought to himself. "We are disposable."

The grand jury convened twice a week to interview nearly one hundred witnesses and review some one thousand six hundred exhibits.

The investigation produced six thousand pages of transcript. The jurors were sworn to secrecy. No one, not even spouses or family, could know about their work. All nineteen worked diligently, with only three unable to finish their terms. They examined records of transmission line inspections to see whether there had been any change in methods or frequency over time and reviewed emails between employees, monitoring for conversations about budget constraints. They were looking for any indication that employees had known about risks on the Caribou-Palermo.

Within a year, the jurors had concluded why PG&E had failed to recognize that a hook on a century-old transmission tower was dangerously close to breaking. It wasn't just the Caribou-Palermo. The company, over the course of decades, had worked to reduce costs associated with inspections, creating a system-wide problem that would take years to fix.

The grand jury had subpoenaed nearly every relevant historical record PG&E still possessed. The company's inspection and patrol policies were documented in "bulletins," and PG&E employees had rummaged around in search of the various iterations that had been produced over time. The oldest one they found, called "Routine Patrolling and Inspection of Transmission Lines," dated to 1987. It required a line such as the Caribou-Palermo to be patrolled three times each year, once by an inspector walking beneath it and twice by helicopter. The policy also required inspectors to climb 5 percent of its towers each year. That way, each structure would be inspected at close range at least once every twenty years, for wear that happened slowly over time.

Similar to pipeline inspection methods, tower inspection methods vary in what they can effectively assess. By walking a transmission line, an inspector can see whether lines appear damaged or are too close to the ground. By flying one, inspectors can get a better look at the tops of the towers and the way the wires connect to them. Climbing the towers, however, has historically been one of the most surefire ways to closely examine the tiny hooks and clamps holding up the conductors.

The 1987 bulletin contained an appendix with a list of conditions for inspectors to check during their patrols. It required them to look for "worn hardware and connectors," including hooks supporting insulator strings. There was a reason for that explicit instruction. That year, in 1987, PG&E's engineering department conducted a series of lab tests on hooks holding up another 115-kilovolt line, similar to ones found along the Caribou-Palermo. One of the hooks was visibly worn, a condition the engineers noted was "possibly caused by the insulator string swinging in the wind over a period of time." The manufacturer's assessment indicated that the hook should be able to hold as much as 30,000 pounds of tension. It failed at 11,500 pounds. PG&E supervisors, aware of the potential hazard, distributed photos of the worn hooks to troublemen in every region for training purposes.

The 1987 policy remained in place until 1995, as the company slashed head count and cut spending in preparation for deregulation. The new guidelines substantially reduced inspection frequency for transmission lines. Requirements for the Caribou-Palermo changed from three inspections a year to one ground inspection and one helicopter inspection every two years. Climbing was no longer routinely required.

The 1995 policy was revised in 2005, the year Peter Darbee became CEO and launched the transformation effort. Under the new policy, the Caribou-Palermo would be inspected from the ground only once every five years. In the off years, it would be quickly patrolled from the air.

In its search for records on the Caribou-Palermo, PG&E found no inspection or patrol logs created prior to 2001. As far as employees
could tell, the line was subject to five ground inspections between
2001 and 2014, a schedule reflective of the 2005 change in policy. The
company had no record of anyone climbing any of the towers, save for
ten of them as it considered a relocation project in 2010.

Ramsey and Noel interviewed anyone who had ever flown the
line by helicopter, the means of patrol during years without inspections. One of them, a former troubleman who had done so prior to
2001, said that flying the Caribou-Palermo typically took at least a
day, and sometimes as long as a day and a half. That was the amount
of time required to examine every tower.

In 2011, inspectors took just over three hours to fly the Caribou-
Palermo. In 2013, they flew the Caribou-Palermo, as well as two other
nearby lines, in about eight hours. In 2015, they flew the Caribou-
Palermo and seven other lines in six hours. And so it went. In 2018,
in the months prior to the Camp Fire, they flew the Caribou-Palermo
and five other lines in five and a half hours.

A retired PG&E employee who spent over thirty years in the
transmission division reviewed the flight records and determined
that the patrols weren't really patrols at all, but "flybys." Another
said that such patrols were only meant to confirm that the towers
were "standing upright." Both agreed: there was no way to evaluate
the state of the tiny hooks and hanger plates from a helicopter speed-
ing fast along any given line.

It came down to money. Each year, the company's finance department determined how much it should spend on inspections and

patrols in each of the transmission maintenance divisions. This sort of work was distinctly different from capital investments on which the company could earn a return. Maintenance money came out of the expense budget. Minimizing those costs—by either reducing them or attempting to shift them into the capital budget—had the potential to benefit shareholders, and the grand jury uncovered evidence that the company had for years attempted to do both.

The dynamic was similar to the simple example that Dr. David Hofmann, the University of North Carolina professor, had offered in his presentation to the CPUC some years earlier. If a budget manager invests $1.00 in capital improvements, he gets $1.20 back. Spending $1.00 on maintenance, meanwhile, is just $1.00 out the door, generally with no immediate reward. If the manager spends just $0.80 on maintenance, then he has $1.20 to invest as capital, with the prospect of a larger return and no obvious consequence. Hofmann had aptly described it as "a slow drift into failure."

—

Year after year, the towers along the Caribou-Palermo went unclimbed, even though engineers were aware that the line had deteriorated to the point of danger. In 2007, PG&E introduced the Deteriorated Transmission Equipment Replacement Program. An engineer used it to request $800,000 to replace several towers along the Caribou-Palermo. In an internal email, he noted that since 2002, there had been five equipment-related failures on the line. Some involved the wires themselves, which had lost strength with time. "It is very time consuming and costly to correct any failures that occur in this dilapidated line section, especially during the winter months when failures are more likely," he wrote. "The probability of that failure is imminent due to the age of both the towers and the conductor." The company authorized $200,000 for the project. The engineer got to work studying sites for the new towers and building a road to better access them.

The following year, in September, the engineer's concerns bore out. One of the line's conductors failed, showering the ground with sparks. A fire ignited and grew to consume five acres of the Plumas National Forest. The tower replacement project wouldn't have prevented it. But it was a sign of more problems to come.

The engineer kept at it, exceeding his $200,000 limit. The overruns were supposed to be charged to the expense budget, but the grand jury found evidence that he faced pressure to find ways to capitalize the additional costs. Two years later, before construction started, the project got axed. He tried to push back. "We have already notified FERC of the project and it will not look good if towers we have identified as deteriorated fall over in the canyon because we did not perform the work due to funding," he wrote in an email, referring to the Federal Energy Regulatory Commission. His supervisors didn't reinstate the project, even when, at the end of December 2012, five of the line's towers collapsed completely.

The engineer kept reminding others that the transmission system in the Feather River Canyon was deteriorating. In 2016, he wrote to his supervisor flagging the need to identify the life expectancies of the old transmission lines the company had purchased from other utilities. The ones it acquired in its merge with Great Western were among the oldest in the system.

"Caribou-Palermo (old Caribou-Golden Gate) for example . . . Built roughly in 1907. This line is in a very remote area. Access is extremely limited. Conductor was deemed annealed several years back. Line has tons of splices in it. Some spans have 5 splices within said span," the engineer wrote, referring to strength loss and potential weak spots as a result of damage repair. "Most of the upper line section is subject to rockslides that have taken this line out in the past. Restoration time is lengthy."

But the transmission division was still under pressure to cut costs. That same year, a finance manager emailed a transmission manager about an upcoming meeting. "The purpose of this meeting is to obtain Leadership guidance on *which* items to pursue and *when*," the email read. "This input is important given the Expense reduction pressure being pushed down on Transmission Operations for 2017."

———

Meanwhile, other projects secured funding without issue. Many of them were located around the Bay Area, where transmission line outages could leave tens of thousands of people without power. The grand jury found that in deciding where money should be spent, company supervisors focused largely on the lines that served the most people. They unearthed several internal memos that characterized the company's maintenance approach as what's known as "run to failure," or replacing parts only after they break. The company took a more proactive approach with targeted improvements on poorly performing lines—namely those that did the most to boost reliability metrics.

A former transmission supervisor was asked if he had noticed a difference in the way lines were inspected and maintained in places like the Sierra foothills. "We're kind of out-of-sight, out of mind up there," he responded. "But if something flips the screen down there [in the Bay Area] they get a lot of attention."

The email that had caused Ramsey's heart to sink captured that mindset. When it came to the Caribou-Palermo, the engineer tasked with assessing it knew that the likelihood of structure failure was high, but didn't flag it as a concern because such failure would leave less than a thousand customers in the dark. The same was true of dozens of old lines crisscrossing the forested Sierra foothills. Great Western's system was one of several built decades earlier to carry power from other hydroelectric projects deep in the mountains.

In the days after the Camp Fire, PG&E sent inspectors to climb towers on other nearby lines built around the same time as the Caribou-Palermo, most for the first time in years, if not decades. Only then did they realize the extent of the problem. Dozens of hooks and other hardware pieces were on the verge of failure, to say nothing of the wires and the structures themselves. Many of the company's engineers and inspectors had known there were hazards throughout the system that could result in fires. But none had documented the extent of them, even as a prolonged drought and an army of bark beetles sapped the life of the forests, heightening the consequence of a single spark.

Ramsey and Noel pondered the question of criminal liability. By their estimation, the wear on the hook that broke had been visible for at least fifty years. But because no one apparently observed it, no single person was negligent. And none of the inspectors and engineers who had flagged the Caribou-Palermo's safety problems bore direct responsibility for its upkeep. The risk management process at PG&E involved dozens of executives and midlevel employees. Ramsey called it decision-making "by committee." Because the company's actions were determined by such a large number of people, no one bore individual responsibility for the consequences.

On top of that, the prosecutors found, the company did a poor job sharing information between departments and management groups. PG&E's tower division was tasked with maintaining the tall steel structures, while its line division looked after the wires. Neither paid attention to the hooks, thinking it was the other division's responsibility to do so. In Ramsey's words, the hooks became orphans, with no department charged with maintaining them.

As Noel had suspected, PG&E bore criminal liability for the Camp Fire and the San Bruno explosion for similar reasons. Just as PG&E's pipeline engineers hadn't known of the exact problem with the seam that ruptured, its electrical engineers hadn't been aware of just how worn the hook had become. In both instances, that lack of knowledge stemmed from the company's poor inspection practices, the thoroughness of which had been reduced over time. But employees in both the gas and electric divisions held other knowledge about risks throughout their systems, information that should have prompted them to do more to assess the hazards. Instead, through their interactions with each other, they ignored the problems or made them worse. That, Ramsey determined, amounted to reckless negligence, the standard needed to indict PG&E itself on eighty-four counts of involuntary manslaughter for its role in the Camp Fire.

Like their San Bruno counterparts, Ramsey and Noel were hamstrung in their ability to levy a meaningful penalty against the company. The statutory maximum fine for the crime was $3.48 million, an amount that would bankrupt most individuals but would have next to no effect on a company the size of PG&E. So Ramsey attempted to negotiate a plea bargain. In the spring of 2020, he met with the company's lawyers to offer a deal: he would charge the company with arson, rather than manslaughter, if it agreed to pay $200 million in criminal and civil penalties. The money would have gone a long way in Butte County, where a third of the tax base had been destroyed by the Camp Fire. To his surprise, the attorneys rebuffed the offer. A higher penalty, they said, could spook investors, particularly those that had agreed to backstop the equity raise at a time when the pandemic had cast global markets into turmoil. Wall Street could apparently overlook one of the most serious of felonies, so long as PG&E stayed on track to exit bankruptcy.

On the morning of June 16, 2020, Bill Johnson donned a dark suit and arrived at Butte County Superior Court in the valley below Paradise. He had been CEO of PG&E for just over a year. He entered the courthouse flanked by two attorneys who sat beside him before the dais in a dimly lit room. A projector screen to their right showed seven rows of twelve thumbnails—the faces of the Camp Fire victims.

Johnson squared his broad shoulders as he stood to face the judge. He waived PG&E's right to a trial. The judge straightened the indictment papers and began reading aloud.

"On behalf of PG&E, how do you plead to violation of Penal Code Section 452: unlawfully causing a fire, as a felony, as alleged in count one of the indictment?" the judge asked.

"Guilty, your honor," Johnson replied.

Then the judge began reading the names of the dead as their photos appeared, one by one, on the screen.

Count two: Joyce Acheson, a sixty-eight-year-old woman who didn't pick up her sisters' frantic warning calls.

"Guilty, your honor," Johnson said, eyes locked on her photo.

Count forty-eight: Sarah Magnuson, who, at seventy-five, had wrapped herself in a wet carpet and sheltered in the bathtub.

"Guilty, your honor," Johnson said.

Count eighty-five: David Young, dead at sixty-nine after crashing his minivan while trying to escape.

"Guilty, your honor," Johnson said.

It took Johnson a half hour to finish the plea. With that, PG&E was convicted of the deadliest corporate crime in American history.

—

As Johnson began his plea, another set of attorneys gathered in a different courtroom in front of a different judge to hear the most consequential decision in PG&E's bankruptcy case. Dennis Montali was about to clear the way for the company's emergence. The company would end the process with more debt than when it started; its shareholders, led by Knighthead and Abrams, would lose very little, having successfully fended off the bondholder plan that offered fire victims more cash at the expense of existing equity. Baupost, the hedge fund, and other companies in the insurance group would collect their $11 billion in cash and profit handsomely on the claims they had purchased at a discount. And the $13.5 billion trust established for the fire victims would be funded half with cash, half with some 470 million shares in the company. Lawyers hadn't succeeded in guaranteeing that stock would be worth exactly $6.75 billion in the race toward the June 30 deadline. Its value would depend on the company's share price upon wrapping up the case.

Unlike PG&E's first spin through bankruptcy, the restructuring did not restore the company's investment-grade credit rating, given that it had emerged even more leveraged than it was at the outset. All three of the big ratings agencies—S&P Global, Moody's, and Fitch—assigned junk ratings to its bonds. For PG&E, borrowing money was going to become more expensive at a time when it had billions of dollars of investments to make.

When it came to equity, the company was struggling to raise roughly $9 billion. Investors including Knighthead and Abrams had agreed to "backstop" the equity raise by purchasing shares at about $7.50 each—substantially less than the price at which shares had traded for most of the prior year, making it easier for the investors to sell at a profit later. The backstop was a last resort, for selling shares at such a low price would depress their overall value. The alternative was to try to sell the shares on the open market, but it was a difficult proposition—PG&E was a risky investment.

About a week earlier, the company had announced a plan. It had agreed to sell about $3.25 billion in shares at a discount to a different group of hedge funds that would serve as a slightly cheaper alternative to the backstop investors. The group had agreed to pay up to $10.50 a share, more than the backstop price but still less than what PG&E, in an ideal world, should have been able to fetch. PG&E would attempt to sell the remaining $5.75 billion through conventional offerings. The company's share price fell on the news, which had indicated to Wall Street that its market value wouldn't quickly rebound.

At the bankruptcy hearing, an attorney for the company appeared before Montali to ask him to approve the terms of the deal, which he said would allow the company to forgo the backstop and sell shares at a price that would better preserve the value of the equity, including that issued to the fire victims' trust.

Will Abrams was deeply skeptical. How was it, he thought, that the company could offer discounted shares to all kinds of investors and expect the victims to benefit? He had arrived prepared for an uphill battle. It was his last chance to make his case for why Montali should reject the reorganization plan. The latest deal to sell shares, in Abrams's view, was just another indication of the risks victims faced in accepting a settlement that afforded half their compensation in stock.

"It is likely to take five or six years for the value of the stock to rebound from this," Abrams told Montali.

"But, Mr. Abrams, what do you think would happen to the stock today if I disapprove this motion?" Montali asked. He pointed out that the financing proposal made it less likely that the company would have to resort to the backstop, which would further depress the company's share price.

"Your Honor, this still is a tale of woe," Abrams said. "This still exposes victims to another summer, likely another few summers, of risk."

Montali, voice edging toward exasperation, reminded Abrams that the state-imposed deadline of June 30, the key to unlocking PG&E's access to the wildfire fund, was two weeks away.

"What is this alternative that you think could be implemented, given these deadlines, if I disapproved the current motion?" Montali asked, noting that whatever investors PG&E had lined up would likely scatter if the company couldn't access the fund. "Nothing could be more of a guarantee to perhaps either kill the plan, which is again, maybe what you'd like, or depress the stock dramatically."

There was no easy or obvious alternative; Abrams understood that. But he couldn't accept the idea of simply pushing the plan through because of an artificial deadline that had fast-tracked negotiations and resulted in a deal that left victims, more than any other claimants, exposed to risk.

"I just flatly disagree with that approach," Abrams said.

Abrams made a few final points before reluctantly resting his case. He had been rebuffed at every turn, each futile attempt to draw attention to the fire victims overshadowed by the demands of those with money and influence: the hedge funds, including Knighthead and Abrams; the personal injury lawyers led by Mikal Watts; even Governor Gavin Newsom in his determination to get PG&E out of bankruptcy by June 30. Abrams expected the worst: more big fires that would sink the company's share price, depress the value of stock in the trust, and leave PG&E even more financially vulnerable. Wildfire season had just begun.

Within a few hours, Montali reached his decision on whether to approve PG&E's bankruptcy plan.

"Here, to me, the evidence is satisfactory and well established that, despite Mr. Abrams's pessimism, the company is charting a course in the middle of COVID and the depressed economy and bankruptcies popping up everywhere, and still is on track to be able to make this plan effective," Montali said. "It should come as no surprise: I'm going to come to the conclusion that the plan should be confirmed."

In Butte County Superior Court, PG&E's arraignment was a strange affair. It occurred during the height of the pandemic, so a limited number of people were allowed inside the courthouse. Everyone else had to watch it live on YouTube. Noel's children tuned in. As Johnson pleaded guilty to each involuntary manslaughter charge, they began texting back and forth: Is Dad crying? He was. The tobacco-spitting, sailor-mouthed prosecutor had been brought to tears in hearing each name.

PG&E's attorneys returned to the courtroom for each of the next two days. Johnson, who had decided to resign upon seeing the company through bankruptcy, never reappeared. In his place was a board member who would serve as the company's interim chief executive after Johnson's departure. One by one, family members of the victims made statements. Some were angry, their voices sharp with hostility. Others cried as they spoke. The attorneys sat stoically, their faces shrouded by KN95 masks.

After the arraignment, the grand jury had a picnic. The jurors were sad, but proud. It was difficult to feel that any justice had been served when no individuals were made to answer for the crime. But they had done their level best. Their work had made history. Very few corporations have been convicted on homicide charges. In 1978, an Indiana grand jury indicted Ford Motor Company on three counts of reckless homicide after a Ford Pinto exploded during a crash, killing its three occupants, but the company was later acquitted. BP pleaded guilty to eleven felony counts of "seaman's manslaughter" and admitted negligence after the 2010 Deepwater Horizon explosion and oil spill in the Gulf of Mexico. PG&E's indictment had been unparalleled.

More than a year earlier, just weeks after the Camp Fire, Judge William Alsup had inquired about whether PG&E would have violated its probation if it had recklessly operated any of its power lines. In their response, the prosecutors who led the San Bruno investigation told the judge that the nature of the violation would depend on PG&E's state of mind. Had the company acted with criminal negligence, it could face involuntary manslaughter charges, they said. And had it acted with malice, it could have committed murder.

The prosecutors had raised the prospect of murder purely as a hypothetical. Of course PG&E executives, managers, engineers, and inspectors hadn't allowed a transmission line to fail with the intent to kill eighty-four people. But Noel had been thinking. There had been a case, decided in 1981 by the Supreme Court of California, involving a man named Robert Watson, who had killed two people while driving drunk. Watson had been convicted of DUI charges several times before. So prosecutors charged him with not only vehicular manslaughter but also second-degree murder, arguing that his prior convictions had made him well aware of the danger of his behavior. Because he drove drunk again, with that knowledge, he had acted with a sort of malice, the prosecutors argued. First-degree murder requires evidence that a person expressly intended to kill another. Second-degree murder, meanwhile, only requires evidence of implied malice—willfully acting with a conscious disregard for human life. Watson was convicted on all charges. His attorneys failed in their appeal. The precedent became known as a Watson murder.

To Noel, it seemed as though the principle could apply squarely to PG&E. The company had been twice convicted on charges of failing to safely maintain its pipelines and power lines. Those convictions, he thought, were more than enough to make the company aware of the risks throughout its system. So, if another wire fell or another pipeline exploded with deadly consequences, he saw no reason why the company shouldn't face second-degree murder charges.

In court, after the last of the victims' family members had stumbled through a tearful statement, Ramsey stood up to give his closing remarks. He invoked the case, noting that it had led to what's known as a Watson warning, an admonition given to every person who applies for a California driver's license. The warning advises that driving under the influence of alcohol or drugs or both is extremely dangerous to human life, and that repeat offenders could face second-degree murder charges. Ramsey proposed a Watson warning for PG&E executives, one that stated the danger of operating an inherently dangerous business "under the influence of greed or profit or both" and advised that another failure could lead to murder charges—potentially against company executives, given their awareness of the circumstances that resulted in the company's conviction.

"Knowledge is power," Ramsey said. "PG&E and its executives should—and I believe have—absorbed the knowledge of how its culture has led to reckless behavior to cut corners to maximize profit, and they need to reform, or they will reap the ill wind of a substantially personal prosecution."

Noel had printed out a summary of the Watson case and framed it. He planned to give it to Johnson. But he never got the chance. Two weeks after leaving Butte County Superior Court, Johnson left PG&E.

EPILOGUE

On July 13, 2021, a tree fell on a PG&E distribution line about twenty-five miles northeast of Paradise, deep within the Feather River Canyon. Two fuses blew. PG&E recorded an outage on the line around 7:00 a.m. The company didn't flag it as an emergency. Hours later, around 11:00 a.m., a troubleman was dispatched to check on the cause. It took him hours longer to reach the line in question, which was accessible only by a remote road winding through the canyon. A bridge was closed for maintenance work, which caused a long delay. He finally arrived on the scene to find a tree leaning on a line, a small fire burning beneath it. He attempted to put it out himself, lugging two extinguishers down a steep, scrubby hill. Cal Fire arrived to help. When it looked as though the blaze were under control, the troubleman called his dispatcher to report that he'd located the two blown fuses, and opened a third to stop the flow of current.

"There's a tree on the line that started a fire," the troubleman told the dispatcher.

"Oh, my goodness," the dispatcher responded.

"Our Cal Fire liaison was up here, all of that. The fire was only, I'm guessing, about three acres," he said. "They said it was at two acres and pretty much contained."

"There's no wire down or anything like that? Everything's good?" the dispatcher asked.

"Yeah."

"I'll just say overhead conductor normal. Okay, awesome," the dispatcher said.

"And I guess when they give us the green light to go back up there, we'll just kind of do a little safety look at about two or three poles and reenergize it," the troubleman said.

"Okay, sounds good. We'll probably talk sometime tomorrow then, huh?" the dispatcher asked.

"Yeah, I would figure so," the troubleman responded.

But the fire, which would become known as the Dixie Fire, wasn't contained. It spread through the canyon with devastating speed. California hadn't seen rain in months. Trees and grass were tinder dry. About 85 percent of the state faced extreme drought conditions, up from less than 3 percent at the same time the prior year. Worse still, it wasn't especially windy on the day the fire ignited. It propelled itself, with no shortage of fuel, blazing out of control for nearly three months. The fire spread up through the canyon to surround the perimeter of Lake Almanor, the reservoir that completed Great Western's power generation system more than one hundred years earlier. It ultimately consumed nearly a million acres, making it the second largest wildfire in California history.

The Dixie Fire weighed on PG&E's share price, which had scarcely rebounded since its emergence from bankruptcy. That had consequence for victims of earlier fires, the majority of whom were still waiting on compensation years after their homes were destroyed. In order for the value of the stock in the trust to reach $6.75 billion, shares in PG&E had to trade at $14. When the trust was funded, in July 2020, they were trading at $9, putting their total value at about $4.2 billion. A year later, that had scarcely changed. The trust's assets were worth a total of about $10 billion, $3.5 billion less than the nominal value of the settlement Mikal Watts and others had pushed so hard to secure.

As the Dixie Fire blazed, the trust hadn't yet sold a single share as it grappled with challenges in estimating claims, as well as unforeseen tax issues. It had made partial payments to less than 5 percent of victims and distributed only about $600 million in cash, nearly half of which amounted to small preliminary payments to those facing acute hardships. A retired judge overseeing the trust estimated it would be months before the trust could begin liquidating the stock. And because it held such a large stake in the company, it would have to do so slowly. A rapid sell-off would sink the share price.

Will Abrams hadn't given up the fight, even though it was too late to reverse the outcome of the victims' settlement. In the wake of the Dixie Fire, he filed another motion in bankruptcy court, where PG&E was resolving a few final issues. The fire, he wrote, should come as no surprise, nor should the trust's performance. He had said it repeatedly: victims of past fires had been left to bear the risk of future ones.

"These eventualities were recognized as highly likely given that nothing within the bankruptcy proceeding restructured the company around safety objectives," he wrote. "Just outcomes for the public and victims have been undermined by the parties in this case."

Shortly after the Dixie Fire ignited, Mike Ramsey and Marc Noel started another investigation. This time, they would have to learn about electric distribution. The question was how PG&E had maintained the small line—and whether the company should have removed the tree that fell on it. The fire was burning frighteningly close. Ramsey, at one point, began packing out of concern that he and his wife would have to flee. For Noel, the pride of having successfully prosecuted PG&E had since been clouded by the realization that its conviction was little more than a symbolic victory. "We made history," he said as he prepared to dig into the first batch of PG&E inspection documents he had requested in the Dixie Fire investigation. "But we didn't change a fucking thing." He had taped a new sign to the wall of his office, right next to the light switch. It was a printout of the PG&E logo that read "Pray to God & Evacuate."

After Bill Johnson stepped down as CEO, PG&E's board searched for a replacement who would stay long enough to make real changes. It was a very hard sell. California had been consumed by the most severe drought in years, and conditions were getting worse each month. PG&E had seen an exodus of employees and executives, many of whom jumped ship during the bankruptcy. The company had lost the trust of its customers and regulators. And its power lines still posed serious risks, even as it tried to install more safeguards and cut more trees. The board ultimately chose a woman named Patti Poppe, a career utility executive and lifelong midwesterner who saw the PG&E job as one that would define her career. She arrived in January 2021.

It soon became clear that 2021 would be one of climate extremes. In February, a freak winter storm settled over Texas, plunging temperatures well below freezing. Power plants of all kinds ceased to function as residents across the state cranked the heat. Electricity demand threatened to exceed supply. The Texas grid operator, facing the prospect of total system breakdown, ordered the state's utilities to cut power to millions of people in order to maintain balance. Some shivered in darkness for four days. Hundreds died in the cold.

Summer came hot and dry. The drought wasn't confined to California; it had overtaken the entire arid West. Large reservoirs reached historically low levels, constraining the output of the Hoover Dam and other major hydroelectric facilities. States across the region were at risk of electricity shortages. In California, where conditions were especially acute, the grid operator called on residents to conserve power, to avoid the need for blackouts.

In June, an extreme heat wave blanketed the Pacific Northwest, where summer weather is usually so mild that many homes and businesses don't have air-conditioning. Temperatures reached a record high of 109 degrees in Spokane, Washington. The utility serving the city had never seen such temperatures. Substation equipment sounded heat alarms. The utility took it off-line to avoid serious damage to the system, cutting power to thousands of customers in the process.

Just before Labor Day, Hurricane Ida barreled through the Gulf of Mexico and struck Louisiana and Mississippi with brutal force, knocking out more than two hundred miles of transmission lines operated by the regional utility, including all eight of the lines serving New Orleans. About nine hundred thousand of the utility's customers lost power in the sweltering heat. It took the company more than two weeks to restore it.

Elsewhere, weather patterns were just as strange. Minnesota saw a surge in wildfires as a severe drought shrank its many lakes. A barrage of summer storms battered Michigan with unusual intensity. A rare spate of tornadoes whipped through the several mid-Atlantic states. Many people recognized the drumbeat of disaster for what it was: proof that climate change was no longer an abstract risk. Its acute consequences, first most evident in California, and then the rest of the West, were manifest throughout the country.

In one way or another, the anomalous weather exposed the vulnerability of the grid. Once an inconspicuous machine, its failures were becoming more obvious, and more consequential. The grid had been built to withstand climate patterns of the past, and those patterns were changing fast at a time when electricity had never been more critical. Every utility would soon face the same question: How should its system change to account for future risk?

Upon arriving at PG&E, Poppe got out in the field to see the extent of the problems for herself. She was floored by the sheer number of trees the company would have to remove in a constant game of catch-up. Meanwhile, the climate would continue to change. While walking through the forest near the Napa County town of Calistoga, she snapped a photo. Employees in yellow vests walked among several large trees marked for removal, all to reduce the risk of fire and the need for wind-driven power shutoffs. She set the photo as her lock-screen wallpaper. It was time, she thought, to reassess the company's strategy.

In June, under Poppe's direction, employees across multiple divisions convened to discuss a wholly different approach to managing power lines. They considered whether the company could do more to bury them in a process known as undergrounding. The reason was simple: underground wires can't start fires. But the burial process is complex, and far more expensive than building wires overhead. For those reasons, very few American utility companies operate large sections of their systems underground.

PG&E had started on a selective undergrounding program in Paradise, where it had committed to burying all of its distribution lines as the town was rebuilt. (In a twist of irony, the line that ignited the Dixie Fire had been targeted for burial.) But the company had stopped short of expanding the program to other areas at high risk of fire, arguing that doing so at scale would be prohibitively expensive.

The start of the Dixie Fire that July left Poppe with a feeling of foreboding. What looked like a healthy Douglas fir had fallen onto a power line on a day when there was scarcely any wind to push it over. It was, in some ways, just an unfortunate accident. But there was almost nothing Poppe could say in the company's defense. How could she stand up and repeat the same tired apology, the one that each of her predecessors had been spinning out since the mid-1990s?

On July 21, a few days after the fire ignited, Poppe traveled to the Butte County city of Chico. A cloud of smoke was visible in the distance, growing larger as the fire spread. Poppe, whose shoulder-length brown hair was streaked with gray, put on a fluorescent safety vest and approached a microphone standing beside a PG&E truck.

"That tree that fell on our line is one of eight million trees in strike distance of our lines," she said, her hazel eyes bright with emotion. "This is an extraordinary problem."

She explained what PG&E was trying to do about it. That year alone, it planned to remove three hundred thousand trees and trim more than a million of them. But it wasn't enough, and likely never would be. The company, Poppe said, would bury ten thousand miles of distribution lines, more than any other utility in the nation.

It was a stunning announcement, and a risky one. She had just committed the company to an unprecedented engineering feat at a time when it had barely begun fleshing out the details. It hadn't so much as set a time frame for completing the work, much less figured out which circuits should go underground. Its employees hadn't even had any substantive conversations with California Public Utilities Commission, which would have to vet the cost and feasibility of the work.

The announcement left the board of directors reeling—Poppe had surprised them the night before with her intent to go public with the plan. Some were upset, out of concern that she had vastly overpromised what the company could deliver. But Poppe felt she had no other choice. The company couldn't prevent more trees from falling on more lines. Undergrounding was essentially the only way to eliminate that risk on any given circuit. It would reduce the need for perpetual spending on cutting trees and redesigning the overhead system. And it would allow the company to keep more of its lines running when the Diablo winds made their seasonal return.

The challenges were large enough to raise serious questions about whether PG&E could bury even half the number of lines it was proposing. Undergrounding would require extensive permitting, complex engineering, and a large amount of labor. It wouldn't address risks to transmission lines, which are even more difficult to bury than distribution lines. And it would be hugely expensive. Already, the company was asking the CPUC to authorize $7.4 billion in new investments through 2026. Customer bills, already driven higher by prior spending on wildfire mitigation and renewable energy projects, were projected to rise 5 percent each year.

Across the country, other utilities were closely watching PG&E. Its undergrounding plan proposed one of the most significant system overhauls ever undertaken in the industry, and its success could help drive down the costs of burying wires. Undergrounding was emerging as a boardroom topic elsewhere as executives and regulators recognized the need to prepare for more extreme conditions. But redesigning a decades-old system would require billions of dollars and years of work. It meant balancing the interests of shareholders and customers while convincing regulators of the need to make large capital investments in the system. Action was necessary, and it would inevitably come at a cost. Poppe estimated PG&E's undergrounding work would cost $20 billion. It was a huge sum, and very likely a conservative one. But maybe it was simply the price the company and its customers would have to pay. The price of aging infrastructure and systemic neglect. The price of providing a critical service in a changing climate. The price of preventing more destruction, and more deaths.

"We know that we have long argued that undergrounding was too expensive," Poppe said. "This is where we say it's too expensive not to underground."

Acknowledgments

The Camp Fire started just three days after I joined *The Wall Street Journal*. One of the biggest stories in corporate America had fallen squarely on my lap, and I hardly knew a thing about utilities, much less PG&E. Enter Russell Gold and Rebecca Smith, two veteran reporters who were extraordinarily generous with their knowledge as we unraveled one of the year's most competitive and complicated stories. The series of articles we produced exists in large part because of Russell and Becky, and serves as the foundation of this book. Russell thought to comb through mind-numbing FERC documents and file public records requests with the federal forest service in search of transmission-spending details. And Becky drew on an invaluable depth of institutional knowledge, as well as relationships with several former executives and employees built during her many years on the utilities beat. My deepest thanks to both of you.

Our stories would not have been half as good without our incredible editors. Miguel Bustillo and Jenn Forsyth elevated our work at every step. Steve Yoder, Alex Martin, and Matthew Rose took it even higher. Thank you all. Thanks as well to Jamie Heller, Matt Murray, and Karen Pensiero for your support of both our coverage and this book.

Our PG&E coverage received significant recognition, including a Gerald Loeb Award and a Pulitzer finalist nomination. The packages submitted to each of those prize committees contained stunning work by very talented visuals editors and photographers. Thanks to Renée Rigdon, Yaryna Serkez, and Dave Cole. And thanks to Erin Ailworth for a deft and emotional account of the effort to rebuild Paradise, the piece of the Pulitzer package that did the most to humanize the consequences of PG&E's negligence.

Other *Wall Street Journal* reporters played significant roles in covering the PG&E fallout. Special thanks to Peg Brickley and Andrew Scurria for tireless coverage of the company's slog through Chapter 11. I consulted it often in writing.

My agent, Laura Usselman, was the first to recognize this book's potential. Thank you for your guidance and encouragement. Without you, I never would have had faith in the wide appeal of such a complex narrative. I feel so fortunate to have worked with Portfolio, which invested in bringing this book to fruition. Thank you to Adrian Zackheim and the rest of the team for supporting my vision for it. My editor, Trish Daly, made the narrative immeasurably better with thoughtful questions and suggestions. Thank you.

A nonfiction book, like any work of narrative journalism, is only as good as its sources. Thank you to the dozens of former employees, regulators, and consultants who talked with me about PG&E. Your stories, memories, and perspectives added so much context and color.

This is in no way a simple story, and each of you helped me bring it to life with fairness and accuracy.

The history of Great Western is sourced extensively from a wonderful archival trove at the University of California, Riverside. Thanks to archivists there who fulfilled my reproduction requests at the height of the pandemic. Special thanks to Karen Raines for helping determine which materials would be most useful in this pursuit.

I am forever indebted to my friends who read several rough iterations of this story and listened to me talk about PG&E for nearly four years (and counting). Special thanks to Rebecca Elliott, Lindsay Ellis, Karine Semple, Brian Rosenthal, Brad Gowland, Michael Bodley, Rebecca Davis-O'Brien, Chris Matthews, Collin Eaton, and Sean Gryger.

And, of course, thank you to my parents, Rob and Margaret, and my sister, Julianne, for your love and support.

A Note on Sourcing

This book is a work of narrative nonfiction, informed by more than two hundred interviews with former PG&E executives, employees, and consultants, as well as California regulators, politicians, and attorneys, conducted over the course of three years. I interviewed some while reporting on PG&E for *The Wall Street Journal*; much of our work is included in the citations. Others I interviewed specifically for this book.

In reporting for both the *Journal* and this book, I talked to a number of sources who agreed to speak only on background. There were several reasons for this. Some former employees are beholden to non-disparagement agreements and expressed concerns that speaking on the record would constitute a breach. Others were involved in, or feared becoming involved in, the many lawsuits related to the fires, as well as other ongoing legal proceedings. In most instances, I agreed to their terms but made sure to vet their statements and recollections with documents and other people familiar with the circumstances they described. I do not cite anonymous sources in the endnotes.

This narrative is also based on details drawn from thousands of pages of court filings and regulatory documents, as well as internal memos, emails, and presentations produced by PG&E employees. Some company documents are publicly available through court and regulatory databases. Others remain confidential or are held privately. I personally reviewed any confidential or private documents I cite.

Much of the historical information about Great Western Power Company and its hydroelectric development comes from the Water Resources Collections & Archives at the University of California, Riverside. Much of the information about PG&E's early years, meanwhile, is contained in *P. G. and E. of California*, a book written by Charles Coleman in honor of the company's centennial in 1952. This book does not have endnotes, which made it difficult for me to independently verify a number of details about the company's founders. I came across some of the book's details and personal accounts in archived copies of *Pacific Service Magazine*, PG&E's in-house publication in its early days. The rivalry between PG&E and Great Western was documented in newspaper articles retrieved through the digital archive newspapers.com, a truly incredible resource.

The evidence and testimony supporting PG&E's conviction on eighty-four counts of involuntary manslaughter is contained in the transcripts of the grand jury proceeding. The Butte County District Attorney's Office moved to make these transcripts public, but PG&E objected, resulting in a protracted legal battle that had yet to be resolved at the time of writing. Because I did not have access to the grand jury transcripts, I relied on the Butte County district attorney's summary of the investigation. PG&E, in response to an order by US

District Judge William Alsup, took issue with several of the summary's characterizations. I took this response into account. I gave PG&E the opportunity to further clarify its perspective on anything described in the summary. Specifically, I requested an interview with the company's chief risk officer, who oversees all risk management associated with its infrastructure. PG&E declined to make him available.

Every person named in this book had the chance to tell his or her story, no matter what role he or she played in the overall narrative. Some declined or did not respond, and as a result, any thought or feeling attributed to one of those people is based on a characterization by other people familiar with his or her state of mind at the time, or interviews he or she gave to other publications. I used such attributions sparingly.

I gave each of PG&E's former chief executives many opportunities to engage in conversation. None chose to do so.

Notes

Prologue

2 **Because of that decision:** "Evidentiary Hearings in PG&E's Bankruptcy Proceeding (I.19-09-016)," California Public Utilities Commission, Admin Monitor video of proceeding (part 2), 52:30–2:53:52, featuring Abrams's interrogation of Johnson, adminmonitor.com/ca/cpuc/hearing/20200225/2.

2 **Abrams, a management consultant:** Will Abrams, interview with the author.

3 **For Abrams, the irony:** Abrams, interview.

4 **During the ride:** Johnson alluded to this in his response to Abrams; and Mike Ramsey (Butte County district attorney), interview with the author.

4 **The women had huddled:** Butte County District Attorney, *The Camp Fire Public Report: A Summary of the Camp Fire Investigation*, June 16, 2020, 14, butte-county.net/Portals/30/CFReport/PGE-THE-CAMP-FIRE-PUBLIC-REPORT.pdf?ver=2020-06-15-190515-977.

4 **The local sheriff:** Ramsey, interview.

4 **an earnest promise:** Johnson's predecessors each used a version of this line after the 2010 pipeline explosion in San Bruno and the 2017 fire siege. See, for example, Tony Earley, "PG&E Won't Contest Huge Penalty for San Bruno," *San Jose Mercury News*, April 16, 2015; and "PG&E CEO Geisha Williams Meets with Local Leaders Impacted by Unprecedented Wildfires," *PG&E Currents*, October 13, 2017.

One: Indictment

9 **Cell phone out:** Butte County District Attorney, *The Camp Fire Public Report: A Summary of the Camp Fire Investigation*, June 16, 2020, 7, buttecounty.net /Portals/30/CFReport/PGE-THE-CAMP-FIRE-PUBLIC-REPORT.pdf?ver =2020-06-15-190515-977.

9 **The message reached:** *Camp Fire Public Report*, 8.

10 **Residents awoke to emergency:** *Camp Fire Public Report*, 2.

10 **At its fastest, it engulfed:** Emanuella Grinberg, Eliott C. McLaughlin, and Christina Zdanowicz, "A Northern California Fire Is Growing at a Rate of about 80 Football Fields per Minute," CNN Wire, November 9, 2018, lite.cnn .com/en/article/h_4f49047228976f93f6e06bc521e28957.

10 **Dozens of people were left:** The nature of victims' deaths is summarized in *Camp Fire Public Report*, 11–18.

11 **Saw an insulator string:** The troubleman photographed the equipment failure from the helicopter. *Camp Fire Public Report*, 43.

11 **An arc of electricity surged:** *Camp Fire Public Report*, 2.

12 **long for investigators:** *Camp Fire Public Report*, 2.

12 **Millimeter by millimeter:** According to the original schematics of the tower, the hook was fifteen-sixteenths of an inch. *Camp Fire Public Report*, 20.

14 **Ramsey told the Cal Fire:** *Camp Fire Public Report*, 2–3.

14 **Noel had long suspected:** Marc Noel, interview with the author.

14 **show up the local prosecutors:** A good example is PG&E's approach to battling misdemeanor charges following a 1994 wildfire near the tiny town of Rough and Ready. See Tom Nadeau, *Showdown at the Bouzy Rouge: People v. PG&E* (Grass Valley, CA: Comstock Bonanza Press, 1998).

14 **Not long after the Camp:** Noel, interview.

15 **He began reading:** Noel, interview.

15 **Noel took the hook:** Noel, interview.

15 **Noel boarded a county helicopter:** Noel, interview.

15 **Three towers in particular:** Ramsey's team flew the line several times to document towers where the top of the hooks and the top of the hanger holes were substantially larger than three-sixteenths of an inch. *Camp Fire Public Report*, 20.

16 **In December, not long after the fire:** *Camp Fire Public Report*, 44.

16 **The linemen discovered more:** Exponent for PG&E, *PG&E Caribou-Palermo Asset Condition Investigation*, November 1, 2019, 14. Included as Appendix A to *Camp Fire Investigation Report*, California Public Utilities Commission, November 8, 2019, archived as I.19-06-015.

16 **The Caribou-Palermo, a fifty-six-mile:** Stephanie Cimino and Wendy Nettles, "Evaluation of Eleven Transmission Lines Associated with Tower Replacements on the Rio Oso-West Sacramento 115kV and the Brighton-Davis 115kV Transmission Lines in Northern California," June 2017.

16 **The line picked up:** North State Resources Inc., *Upper North Fork Feather River Hydroelectric Project: Draft Environmental Impact Report*, State Water Resources

Control Board, November 2014, 3–5. http://project2105.org/SWRCB/final
_DEIR/nov28_14_swrcb_DEIR.pdf.

18 **Prosecutors alleged the utility:** The utility denied the allegations and agreed
to a $1.1 million settlement.

18 **The leak occurred:** Blade Energy Partners, *Root Cause Analysis of the Uncon-
trolled Release from Aliso Canyon,* May 16, 2019, 203–16.

18 **The engineer overseeing:** National Transportation Safety Board, *Overpressur-
ization of Natural Gas Distribution System, Explosions, and Fires in Merrimack Valley,
Massachusetts,* September 13, 2018, 41.

20 **They returned to one of the three:** Noel, interview; and Mike Ramsey, interview with author.

20 **PG&E had replaced them:** Noel, interview; Ramsey, interview; and *Camp Fire
Public Report,* 21.

20 **Noel was infuriated:** Noel, interview.

20 **Any hint of obstruction:** Noel, interview; and author observation.

20 **He dialed up PG&E's attorneys:** This exchange is based on Noel's and Ramsey's recollections of the conversation.

20 **Noel smiled to himself:** Noel, interview.

21 **Ramsey had to buy:** Ramsey, interview.

21 **It had been manufactured:** In court documents, PG&E filed plans for the
original Great Western Power transposition towers that included a schematic,
dated October 11, 1912, of an Ohio Brass suspension hook with a raised B on
the right face of the hook. See Russell Gold and Katherine Blunt, "This Old
Metal Hook Could Determine Whether PG&E Committed a Crime," *Wall Street
Journal,* March 8, 2020, wsj.com/articles/this-old-metal-hook-could-determine
-whether-pg-e-committed-a-crime-11583623059

22 **In a plea for funding:** *Camp Fire Public Report,* 34.

22 **The PG&E engineer:** *Camp Fire Public Report,* 35.

23 **"In 1930, PG&E":** *Camp Fire Public Report,* 85.

Two: Building a Monopoly

24 **Howells thought not:** "Man of Vision First to 'See' Huge System," *Oroville
Mercury Register,* April 13, 1937; and Charles M. Coleman, *P. G. and E. of Califor-
nia: The Centennial Story of Pacific Gas and Electric Company; 1852–1952* (New York:
McGraw-Hill, 1952), 211.

24 **Howells, a Hoosier:** It has often been reported that Howells was born in
Ohio, but in fact he was a native of Richmond, Indiana. "Julius Howells," *Richmond Item,* April 27, 1927.

24 **site their tracks across the country:** Willi H. Hager, *Hydraulicians in the USA
1800–2000: A Biographical Dictionary of Leaders in Hydraulic Engineering and Fluid
Mechanics* (United Kingdom: London CRC Press, 2015), 2177.

26 **Henry James Rogers, the:** United States Department of Interior, National
Park Service, National Register of Historic Places Inventory. Nomination form
for Henry J. Rogers House, 625 W. Prospect Ave., August 8, 2014.

26 **Rogers, who also:** Burton Ellinwood Greene, "Electrical Pioneers, A Histori-
    cal Sketch," *Electricity* 3, no. 2, July 27, 1892, 22.

26 **Rogers left the task:** "Kurz Recalls First Days in Light Plant," *Appleton Post-
    Crescent*, September 19, 1932.

26 **Dozens of new:** "Kurz Recalls First Days."

26 **reporter from the *Appleton Crescent*:** "Electric Light," *Appleton Crescent*,
    October 7, 1882.

27 **It would become known:** It appears this term may have been coined in
    France and spread to the United States as the Niagara and Folsom plants were
    built. "A Happy Idea from France," *York Daily*, November 6, 1901.

28 **"meaning of it all reaches":** "Folsom Power Plant," *Sacramento Record-Union*,
    September 10, 1895.

28 **He endeared himself:** "Howells Fiddled in Quincy 30 Years Ago," *Plumas
    National Bulletin*, October 29, 1914.

28 **Howells got in touch:** Schuyler wrote *Reservoirs for Irrigation, Water-Power, and
    Domestic Water-Supply* (New York: John Wiley & Sons, 1901), which for many
    years was a standard work on this subject, according to the Special Collections
    & University Archives, University of California, Riverside.

29 **When it came to water:** Short biography included in description of archival
    papers at the Special Collections & University Archives, University of Califor-
    nia, Riverside; and "James Dix Schuyler," *Engineering News*, April 17, 1913, 786.

29 **In Japan, he fell:** "James Dix Schuyler," 786.

29 **He spent the entire:** "James Dix Schuyler," 786.

29 **"tempting field for electric":** James D. Schuyler, *Report on Water Storage
    and Power Development on North Fork of Feather River, California*, June 30, 1902,
    James Dix Schuyler Papers (WRCA 063), Water Resources Collections and
    Archives, Special Collections & University Archives, University of California,
    Riverside, 38.

29 **Once, he had to:** Roosevelt's full letter is worth a read and can be found in
    "'Teddy' Has Bull for the University," *Oakland Tribune*, March 10, 1910; and
    *Report of the Secretary to the Regents of the University of California for the Year Ending
    June 30, 1910*, University of California Bulletin, series 3, vol. 4, no. 2 (Berkeley:
    University of California, 1910), 66.

30 **"This will by far":** Julius M. Howells to E. T. Earl, March 4, 1902, Schuyler
    Papers.

30 **Prior in time:** The story of Howells's pursuit of the water rights was widely
    recounted in various newspaper articles at the time. For this account, I relied
    primarily on a firsthand recollection by W. H. Spaulding, a PG&E attorney who
    served in Guy Earl's law office, in W. H. Spaulding, "Beginnings of the Great
    Western System—the Story of Lake Almanor," *Pacific Service Magazine*, April
    1931.

31 **Bidwell only just:** Coleman, *P. G. and E. of California*, 214.

31 **That summer, Schuyler:** Schuyler, *Report on Water Storage*.

31 **"The more carefully I have considered":** Schuyler, *Report on Water Storage*, 38.

30 **group of prominent:** Frederick Hall Fowler, *Hydroelectric Power Systems of California and Their Extensions into Oregon and Nevada* (Washington, DC: Government Printing Office, 1923).

32 **Two years later, it:** *Public Utilities Reports* (Rochester, New York: Lawyers Cooperative, 1916), 595.

32 **His father enrolled him:** Coleman, *P. G. and E. of California*, 129.

32 **Tall and brawny:** Archie Rice, "The History of the Nevada Power Plant," *Pacific Service Magazine*, December 1909.

32 **He shoveled coal:** De Sabla's firsthand recollection, as stated in Alfonso Adolphus Tregidgo's obituary. "Alfonso Adolphus Tregidgo," *Pacific Service Magazine*, September 1923.

32 **He found an opportunity:** "Alfonso Adolphus Tregidgo."

33 **At thirteen, he took:** "Famous Captain of Industry Battles for Life," *Santa Cruz Evening News*, May 28, 1928.

33 **A mutual friend introduced:** Coleman, *P. G. and E. of California*, 133.

33 **Martin, a sturdy:** Martin's firsthand recollection, as relayed in "In the Beginnings of Pacific Service," *Pacific Service Magazine*, December 1921.

33 **Martin, who was supposed:** Coleman, *P. G. and E. of California*, 133.

33 **"I'm ready now":** Coleman, *P. G. and E. of California*, 133.

33 **He had traveled:** "In the Beginnings of Pacific Service."

34 **Martin called upon colleagues:** "In the Beginnings of Pacific Service."

34 **Martin hired three:** A. J. Stephens, "Operators Were All Men at Work," *Pacific Service Magazine*, July 1917.

34 **Every now and then:** Stephens, "Operators Were All Men at Work."

34 **As different as:** Archie Rice, "The History of the Nevada Power Plant," *Pacific Service Magazine*, December 1909.

35 **Some within the company:** Coleman, *P. G. and E. of California*, 244.

36 **In a lengthy handwritten:** Spaulding, "Beginnings of the Great Western System."

36 **Howells had known:** Howells to Earl, March 4, 1902, Schuyler Papers.

36 **More than 150 people:** "Many Take Trip to Berry Creek," *Oroville Daily Register*, November 9, 1908.

37 **Great Western had also:** "To Start Work toward Marysville," *Oroville Daily Register*, July 3, 1908.

37 **The oldest villagers:** "To Start Work toward Marysville."

37 **In 1911, it swept:** "Great Western Gets Contract from the State," *Sacramento Bee*, June 12, 1911.

37 **"The Great Western now":** "Western Power Crosses the Bay," *Sacramento Bee*, June 21, 1911.

38 **In September, Great Western's:** "Gas and Power Corporations Plan Combine," *San Francisco Call*, September 30, 1911; and "Plans to Consolidate Big Power Companies," *San Francisco Recorder*, October 2, 1911.

38 **Several of Great Western's:** "Bankers' Coming Revives Talk of Huge Merger," *San Francisco Examiner*, November 21, 1911.

38 **But Great Western was:** "Great Western May Be Gobbled Up by Old Company," *Sacramento Bee*, September 30, 1911.

38 **"is not particularly":** "Fleishhacker Back; Talks on Merger," *Sacramento Bee*, October 19, 1911.

38 **The company proposed acquiring:** "Power Merger Is Postponed," *Plumas Star*, December 22, 1911.

39 **"Not only has the Pacific":** "Stock Dividend Step to Merger," *San Francisco Chronicle*, November 24, 1911.

39 **He began preaching:** Speech reprinted in "Municipal Ownership of Electric Light Plants," *Engineering News and American Railway Journal* 39, no. 24 (June 16, 1898).

40 **"Our business is":** Samuel Insull, "The Obligations of Monopoly Must Be Accepted," in *Central-Station Electric Service*, ed. William Eugene Keily (Chicago: private printing, 1915), 118–22.

40 **The commission would allow:** *Public Utilities Act of California* (San Francisco: Louis Sloss, 1912).

40 **Otherwise, the commission might:** "Decisions and Orders of the State Commissions in the State of California," *Public Service Regulation* 1, no. 8 (January 1912): 538.

40 **The statute reflected:** Chester Rowell, "A Tribute to John Eshleman," *California Outlook* 20, no. 1 (April 1916): 13–14.

40 **"menace to those":** John Eshleman, introduction to *Public Utilities Act of California*, 13.

40 **Eshleman, who went:** Guy C. Earl delivered Eshleman's eulogy after his death in 1916. His speech was printed as "John Eshleman's Career," *California Outlook* 20, no. 2 (May 1916): 38.

41 **He was said to:** "How John M. Eshleman Made Good," *California Outlook* 17, no. 17 (October 24, 1914): 10.

41 **Eshleman recognized the:** John Eshleman, "A Criticism of the Reproduction Theory of Valuation," Proceedings of the Conference on Valuation, *Utilities Magazine* 1, no. 3 (January 1916): 5–12.

41 **"In short, we say":** Eshleman, "Criticism of the Reproduction Theory."

41 **Just weeks after the commission:** "Pacific Gas & Electric Again Attacks Rival," *San Francisco Recorder*, May 29, 1912.

42 **Great Western countered by:** "The Controversy between Rival Power Companies," *Napa Journal*, May 23, 1912.

42 **"It is most important":** See, for example, "Do You Think the Great Western Power Co. Should Be Allowed to Furnish You Electric Service?," *Petaluma Argus-Courier*, May 27, 1912.

42 **"Tell us your needs":** See, for example, "Notice to Consumers," *Berkeley Gazette*, March 30, 1912.

42 **They appeared at the hearings:** "Can We Have Competion [sic]," *Petaluma Daily Courier*, May 29, 1912.

43 **Earl jumped to his feet:** The fight between Cutten and Earl was recounted in several different newspapers, each of which varied to some degree in detail

and level of drama. The quotes and details included here are from "Attorneys Near Fight at Hearing," *Sacramento Bee*, June 7, 1912; and "Cutten and Earl Argue Power Case," *San Francisco Call*, June 7, 1912.

43 **Earl closed his arguments:** "Power Companies Close Their Case," *San Francisco Chronicle*, June 9, 1912.

43 **"If the Great Western wants":** "Power Companies Close Their Case."

44 **It granted Great Western:** "Decisions and Orders of the State Commissions," 538.

44 **It had nearly completed renovating:** "Great Western Power Co and City Electric Co now at 233 Post St.," *San Francisco Examiner*, July 15, 1912.

44 **Come spring of:** "Power Reservoir Is Flooding Fast," *San Francisco Examiner*, May 8, 1914.

45 **That year, in a solicitation:** Carl Jonas Rhodin Papers (WRCA 087), "A 7% Investment Combining Soundness, Safety and Attractive Yield," Water Resources Collections and Archives, Special Collections & University Archives, University of California, Riverside.

45 **He drove a Cartercar:** "Auto Licenses Are Issued for State," *San Francisco Call*, September 14, 1912.

45 **His daughter got married:** "Christine Howells Is Bride," *Oroville Daily Register*, July 3, 1913; and "Nuptial Vows are Exchanged in Garden," *Oakland Tribune*, July 2, 1923.

46 **Black later stated:** Testimony of James B. Black, U.S. Court of Appeals for the Ninth Circuit - 127 F.2d 378 (9th Cir. 1942).

46 **It had articulated its support:** California Railroad Commission Decisions, Volume 38, Decision No. 25587, January 30, 1933.

47 **He became widely known:** "John Eshleman's Career," *California Outlook* 20, no. 2 (May 1916): 38.

48 **PG&E had roughly doubled its capacity:** Congressional Record, Vol. 98, Pt. 11 (Washington, DC: Government Printing Office, 1952), A4254.

48 **This time, he beamed:** General Electric Company, *Phillip Reed Presents Coffin Award to James B. Block* [sic], *Pres. of Pacific Gas and Electric Company*, June 1, 1952, photograph, 8 × 10", Museum of Innovation and Science, Schenectady, NY, Google Arts & Culture, artsandculture.google.com/asset/phillip-reed-presents -coffin-award-to-james-b-block-pres-of-pacific-gas-and-electric-company -general-electric-company/xAHCobOWheDIhA?hl=en.

48 **Decades earlier, at the turn:** Schuyler, *Report on Water Storage*.

49 **Schuyler, the engineer:** Schuyler, *Report on Water Storage*.

49 **lost integrity with age:** W.R. Bean, H.W. Highriter and E.S. Davenport, "Fractures and Microstructures of American Malleable Cast Iron," *Proceedings of the Twenty-fifth Annual Meeting, American Foundrymen's Association*, Volume XXIX (October 1920): 306.

49 **"freed from all tendency":** "Notes of the Industry," *Journal of the American Institute of Electrical Engineers* 45, no. 3 (March 1926): 320.

Three: Death March

50 Peace developed a: Steve Peace, interview with the author; Michael R. Peevey, interview by Roger Eardley-Pryor, February 1, 2019, transcript, 257, Oral History Center, Bancroft Library, University of California, Berkeley; Tony Perry, "Prospects Dark for Deregulation Architect," *Los Angeles Times*, January 3, 2001; and "Hot Temper, Cool Skills," *Los Angeles Times*, December 30, 2002.

51 Peace had buried himself: Christopher Leonard offers an engaging profile of Peace, as well as a brilliant explanation of the California energy crisis, in "Attack of the Killer Electrons!," in *Kochland: The Secret History of Koch Industries and Corporate Power in America* (New York: Simon & Schuster, 2019), 266–293. See also Peevey, interview, 257–58.

51 He was, by his: Peace, interview.

51 Between 1990 and 1993: Federal Deposit Insurance Corporation, "Banking Problems in California," in *History of the Eighties—Lessons for the Future, vol. 1, An Examination of the Banking Crises of the 1980s and Early 1990s* (Washington, DC: FDIC, 1997), 379–418.

51 In 1991, the average: Jeffrey Dasovich, William Meyer, and Virginia A. Coe, *California's Electric Services Industry: Perspectives on the Past, Strategies for the Future* (San Francisco: California Public Utilities Commission, 1993). Also referred to as *The Yellow Book*.

52 Peace set the stakes: Hearing of the Electric Industry Restructuring Conference Committee, 24:00, July 11, 1996, archived as 071196C1-2 in Senate TV/Cal Channel footage retrieved from the California State Archives. Digital reproduction.

53 "It is not too much": Lewis L. Strauss, "Remarks Prepared by Lewis L. Strauss, Chairman, United States Atomic Energy Commission, for Delivery at the Founders' Day Dinner, National Association of Science Writers, on Thursday, September 16, 1954, New York, New York," United States Nuclear Regulatory Commission, nrc.gov/docs/ML1613/ML16131A120.pdf.

53 Nationally, electricity demand: Dasovich, Meyer, and Coe, *California's Electric Services Industry*, 20.

54 Each plant was expected: Dasovich, Meyer, and Coe, *California's Electric Services Industry*, 84; and Donald Woutat and Greg Johnson, "Consumers May Pay Full Tab for San Onofre Unit; Judge Says Management of $4.5-Billion Nuclear Project Was 'Reasonable,'" *Los Angeles Times*, October 25, 1986.

54 Politicians and scientists heralded: "Nuclear 'Beach Ball' Heralds Age of Atom," *Los Angeles Times*, January 29, 1968; and "New Era Opens in A-Power," *Independent Star-News* (Pasadena, CA), January 7, 1968.

54 audience of about four hundred: The president was Joseph Sinnott, as quoted in "Nuclear Power Station Symbolizes Peace Atom," *Times-Advocate* (Escondido, CA), January 7, 1968.

55 Utility customers would soon: Dasovich, Meyer, and Coe, *California's Electric Services Industry*, 31.

55 The utilities signed long-term contracts: Deirdre O'Callaghan and Steve

Greenwald, "PURPA from Coast to Coast: America's Great Electricity Experiment," *Natural Resources & Environment* 10, no. 3 (Winter 1996): 17–21.

57 **And customers were once:** Reginald Sanders, *Analysis of the Projected Electricity Prices to 1995*, August 1985, 1–2.

57 **That didn't sit well:** Sanders, *Analysis of the Projected Electricity Prices*, 1–2; and California Public Utilities Commission, *Order Instituting Rulemaking on the Commission's Proposed Policies Governing Restructuring of California's Electric Services Industry and Reforming Regulation*, April 20, 1994.

58 **The CPUC president at:** California Public Utilities Commission, *Order Instituting Investigation on Commission's Proposed Policies*.

59 **"represent a superior disciplinary":** California Public Utilities Commission, *Order Instituting Investigation on Commission's Proposed Policies*.

59 **As the CPUC debated:** Interview with former PG&E CEO Richard Clarke as relayed in "PG&E Chief Looks Ahead," *San Francisco Examiner*, May 10, 1993; Michael Parrish, "Optimism and Caution at PUC Hearing," *Los Angeles Times*, July 15, 1994; and Patrick Lee, "PUC to Weigh Compromise on Deregulation: Energy: Edison Offers Alternative to Plans Now on Table. Consumer Groups Oppose It," *Los Angeles Times*, August 22, 1995.

60 **"People speak as if":** *Electricity Issues: Hearings before the Subcomm. on Energy and Power of the Comm. on Energy and Commerce*, House of Representatives, 103rd Cong., 2nd session (1994) (statement of Robert D. Glynn Jr., chief operating officer of PG&E).

60 **It announced plans:** "PG&E Announces 3,000 Job Cuts," *United Press International*, February 22, 1993.

60 **It created PG&E Corporation:** The most significant was National Energy Group, which bought power plants in the Northeast and became a large player in the power-trading market; Jonathan Marshall, "PG&E Sees Room to Grow in Australia," *San Francisco Chronicle*, April 18, 1996; and various securities filings.

60 **In truth, Steve Peace:** Peace, interview.

61 **The groans were audible:** Dan Richard, correspondence with the author.

61 **Peace threatened to resolve:** Hearing of the Electric Industry Restructuring Conference Committee, 23:30, August 10, 1996, archived as 081096X1 in Senate TV/Cal Channel footage retrieved from the California State Archives. Digital reproduction.

62 **The next day, Peace told:** Hearing of the Electric Industry Restructuring Conference Committee, 14:00, August 25, 1996, archived as 082596A1 in Senate TV/Cal Channel footage retrieved from the California State Archives. Digital reproduction.

Four: Market Failure

64 **When George Sladoje:** George Sladoje, interview with the author.

64 **But he knew nothing:** Sladoje, interview.

64 **He saw it as:** Sladoje, interview.

67 **Sladoje made some calls:** Sladoje, interview.

67 **The exchange cleared the trades:** Carrie Peyton, "Electric Market Opens for Business," *Sacramento Bee*, April 1, 1998.

67 **All told, $10 million worth:** Michael White, "California's New Electricity Market Opens behind Schedule," Associated Press, April 1, 1998.

67 **Sladoje's team had:** Sladoje, interview.

67 **Peace wrote a letter:** "Restructuring Update: Efficient Power Exchange Benefits All California Electricity Ratepayers," Letter to the Members of the California Legislature from Senate Committee on Energy, Utilities and Communications, May 13, 1998. Letter reviewed by author.

68 **The grid operator's counterpart:** Sladoje, interview.

68 **But the CPUC, recalling:** Michael Peevey and Diane Wittenberg, *California Goes Green: A Roadmap to Climate Leadership* (South Carolina: CreateSpace Independent Publishing Platform); and Steve Issler, *Electricity Restructuring in the United States* (New York: Cambridge University Press).

69 **He wound up studying:** See, for example, Steven Stoft, Timothy N. Belden, Charles A. Goldman, and Steven J. Pickle, "Primer on Electricity Futures and Other Derivatives," LBNL-41098, UC-1321 (Lawrence Berkeley National Laboratory, University of California, January 1998); and Joseph H. Eto, Steven Stoft, and Timothy N. Belden, "The Theory and Practice of Decoupling Utility Revenues from Sales," *Utilities Policy* 6, no. 1 (March 1997): 43–55.

69 **Their training materials were:** Prepared Direct Testimony of Robert F. McCullough on Behalf of Public Utility District No. 1 of Snohomish County, Wash., before the Federal Energy Regulatory Commission, Ex. SNO-58, Docket EL03-180-000.

69 **He was passionate about:** Scott Thurm, Robert Gavin, and Mitchel Benson, "Juice Squeeze: As California Starved for Energy, U.S. Businesses Had a Feast," September 16, 2002; and John R. Wilke and Robert Gavin, "Brazen Trade Marks New Path of Enron Probe," *Wall Street Journal*, October 22, 2002.

69 **But he was also shrewd:** Thurm, Gavin, and Benson, "Juice Squeeze"; and Wilke and Gavin, "Brazen Trade Marks New Path."

70 **One of Belden's earliest:** Wilke and Gavin, "Brazen Trade Marks New Path"; and Prepared Supplemental Testimony of Robert F. McCullough on Behalf of Public Utility District No. 1 of Snohomish County, Wash., before the Federal Energy Regulatory Commission, Ex. SNO-710, Docket EL03-180-000.

70 **It was a huge amount:** Wilke and Gavin, "Brazen Trade Marks New Path."

71 **Sladoje, too, sensed something:** Sladoje, interview; and contemporaneous thoughts quoted in personal email from Gary B. Ackerman (executive director, Western Power Trading Forum) to Tim Belden et al., June 24, 1999.

71 **Sladoje received a form letter:** Sladoje, interview; and letter reviewed by author.

71 **In May, power sold:** Christopher Weare, *The California Electricity Crisis: Causes and Policy Options* (San Francisco: Public Policy Institute of California, 2003), 26.

71 **That number tripled:** William Hogan, "The California Meltdown," *Harvard Magazine*, September-October 2001.

72 **They were prohibited from:** Issler, *Electricity Restructuring in the United States*, 261.

72 **Bulk power cost anywhere:** Issler, *Electricity Restructuring in the United States*, 261.

72 **as much as $50 million per day:** Weare, *California Electricity Crisis*.

72 **Bob Glynn, who had:** Dan Richard, interview with the author.

72 **A number of generators were:** Thurm, Gavin, and Benson, "Juice Squeeze."

73 **In October 2000, a man:** Keith Totzke to Enron PR department, October 17, 2000. The PR department forwarded the email to James Steffes, former vice president of government affairs for Enron Corp., who then passed it to Belden and other traders. "FYI. The attached e-mail came into PR unsolicited. Check out his conclusion. . . . ," Steffes wrote.

74 **He had brought on bankruptcy:** Richard, interview.

74 **He had once been described:** Richard, interview.

74 **But Davis kept his distance:** Richard, interview.

74 **He might have allowed for:** Rebecca Smith, "California's Governor Meets with Chiefs of Edison, PG&E over Rate Increases," *Wall Street Journal*, December 20, 2000.

75 **"I'm pretty sure it landed":** Richard, interview.

75 **But FERC was hesitant:** Peevey and Wittenberg, *California Goes Green*, 64; and Issler, *Electricity Restructuring in the United States*, 286.

76 **US senator Dianne Feinstein:** Rebecca Smith, "Regulators Step In to Ease Price Shocks in California's Deregulated Power Market," *Wall Street Journal*, December 18, 2000.

76 **The incident evoked the irony:** Author's notes from watching *Attack of the Killer Tomatoes!*

76 **Richard drafted a strongly worded:** Letter retained by Richard, reviewed by the author.

77 **On January 16, as the:** Recording was uploaded by Snohomish County Public Utilities District and accessed by author, but has since become unavailable. See Jonathan Peterson, "Tapes Reveal Enron's Power Plant Rigging," *Los Angeles Times*, February 4, 2005; and Timothy Egan, "Tapes Show Enron Arranged Plant Shutdown," *New York Times*, February 4, 2005.

78 **Sladoje felt beaten down:** Sladoje, interview.

78 **At a meeting one Sunday in January:** Sladoje, interview.

78 **In a statement before:** California Power Exchange Corp, No. 2:01-bk-16577, "Supplemental Declarations of George Sladoje and Lynn Miller," Public Access to Court Electronic Records (Bankr. C.D.Ca.).

78 **Glynn told Richard that:** Richard, interview.

79 **He just couldn't tell him:** Richard, interview.

79 **Davis, a thin man:** Gray Davis, "California Energy Crisis," April 5, 2001, Sacramento, CA, C-SPAN broadcast, archived video, 5:51, c-span.org/video /?163609-1/california-energy-crisis.

80 **Richard looked at Glynn:** Richard, interview.

80 **For Davis, it was a:** David Lazarus, "PG&E Files for Bankruptcy; $9 billion in Debt, Firm Abandons Bailout Talks with State," *San Francisco Chronicle*, April 7, 2001.

81 **Generators and traders, once seen:** Thurm, Gavin, and Benson, "Juice Squeeze"; and Rebecca Smith, "Beleaguered Energy Firms Try to Share Pain with Utility Units," *Wall Street Journal*, December 26, 2002.

82 **As the crisis:** Steve Peace, interview with the author.

82 **Up until that point:** Peace, interview.

82 **As he considered:** Rebecca Smith, "Struggle between Utilities and Customers May Affect Future of Energy Deregulation," *Wall Street Journal*, November 27, 2000; and Bill Ainsworth, "Peace's Profile Low in Energy Dilemma," *San Diego Union-Tribune*, January 15, 2001.

82 **"He couldn't be elected dogcatcher":** Perry, "Prospects Dark for Deregulation Architect."

83 **"The judge and jury":** "PG&E Bailout Is Much Too Rich," *San Francisco Chronicle*, December 9, 2003.

Five: Transformation

84 **Foam placards appeared:** Tom Dalzell (former business manager of the International Brotherhood of Electrical Workers Local 1245, PG&E's labor union), interview with the author.

84 **One read STAY THE COURSE!:** Placard sat on author's desk as motivation to finish this book.

84 **Glynn was determined to see:** Dan Richard, interview with the author.

85 **Across the country, it was:** Don Durfee, "The Top Spot: Why More Companies Are Tapping Their Finance Chiefs for CEO," *CFO*, October 7, 2005.

85 **The bankruptcy had ironically:** Hedge funds held between 10 and 15 percent of PG&E Corp. common stock during the Pacific Gas & Electric bankruptcy. Historical ownership statistics accessed via FactSet.

86 **Wilder, at forty-six:** In 2004, *Institutional Investor* named Wilder the best CFO in the utilities industry. TXU Corp., press release, February 23, 2004.

86 **He further cut costs:** Trevor Fetter, Erik Snowberg, and Rebecca M. Henderson, "TXU (A): Powering the Largest Leveraged Buyout in History," case 9-320-064, Harvard Business School, December 8, 2019.

87 **In February 2005, a month:** PG&E Corp., "PG&E Corp. Details Positive Outlook for Investors and Utility Customers," press release, February 25, 2005; and invitation sent to financial analysts. Details about Business Transformation first appeared in Russell Gold, Rebecca Smith, and Katherine Blunt, "Wired to Fail," *Wall Street Journal*, December 28, 2019. Smith was instrumental in our reporting on the Business Transformation effort. She sussed out many details about its objectives and shortcomings through interviews with former executives and through Accenture materials now housed at Cornell University.

87 **He estimated that the company:** Peter Darbee, "PG&E Corporation: A Value Opportunity," (PowerPoint presentation, PG&E investors conference, New York, NY, February 25, 2005), slide 29.

88 **Around the time he pitched:** Peter Darbee, "Remarks before the 2005 Annual Shareholders Meeting" (San Ramon, CA, April 20, 2005), transcript, pge-corp.com/investors/investor_info/presentations/20050420_sp.shtml.

88 **His goal, he said, was:** Various company materials including the *2006 Corporate Sustainability Report.*

88 **They found the word *delighted*:** Gold, Smith, and Blunt, "Wired to Fail"; and Richard, interview.

88 **He promised that the company would:** Darbee, "Remarks before the 2005 Annual Shareholders Meeting."

88 **In June, with:** Peter Darbee, interview by Becky Quick, "CEO Spotlight: PG&E Corp. President & CEO Interview," CNBC, June 6, 2005, transcript.

89 **In the boardroom, Darbee:** Gold, Smith, and Blunt, "Wired to Fail"; and Richard, interview.

90 **"What about safety?":** Richard, interview.

90 **One afternoon, one of Darbee's:** Richard, interview.

90 **A set of elevators:** Richard, interview; and correspondence with a building mechanic.

91 **He didn't feel:** Gold, Smith, and Blunt, "Wired to Fail."

91 **But the turnover:** Gold, Smith, and Blunt, "Wired to Fail."

91 **A cohort of:** Dalzell, interview.

91 **In a 2005:** Accenture, "ESC Benchmarking" (presentation to PG&E, June 8, 2005). Confidential at time of creation; archived at Cornell University.

91 **Accenture proposed the:** Accenture, "ESC Benchmarking."

92 **He was unorthodox:** Observation of the author.

92 **He had made himself:** Observation of the author.

92 **He saw what looked:** Dalzell, interview.

93 **PG&E's rank-and-file workers:** Tom Schneider (chief executive, Restructuring Associates Inc., a consultancy that PG&E hired to work with the unions during Business Transformation), interview with the author.

93 **Some employees, anticipating disaster:** Schneider, interview.

93 **As one consultant put it:** Schneider, interview.

94 **Schwarzenegger, who once demonstrated:** Miguel Bustillo, "Gov. Pulls Hydrogen SUV Out of Thin Air," *Los Angeles Times,* October 23, 2004.

94 **He pulled Peevey aside:** Michael Peevey, interview with the author.

94 **Davis gave him the number:** Peevey, interview.

95 **greenest president:** Peevey, interview.

95 **He said he wanted:** Michael Peevey and Diane Wittenberg, *California Goes Green: A Roadmap to Climate Leadership* (South Carolina: CreateSpace Independent Publishing Platform).

96 **"What'd you think?":** Michael R. Peevey, interview by Roger Eardley-Pryor, February 1, 2019, transcript, 252–53, Oral History Center, Bancroft Library, University of California, Berkeley.

96 **Dan Richard and others:** Richard, interview.

97 **In February 2005, Richard:** Memo reviewed by author.

97 **At the time, Darbee:** Richard, interview.

97 **Though he had never considered:** David Lazarus, "The Education of PG&E's Peter Darbee," *San Francisco Chronicle*, October 18, 2006.

97 **developed a certain respect for Darbee:** Peevey, interview with the author.

97 **The two men began spending:** Peevey, interview.

98 **"If you had asked":** Lazarus, "The Education of PG&E's Peter Darbee."

98 **Schwarzenegger appeared on:** "Save the Planet—or Else," *Newsweek*, April 16, 2007.

98 **The following month, *Vanity Fair*:** Art Streiber, "Portraits of Ecological Heroes," *Vanity Fair*, May 2007.

98 **Between 2007 and 2010, PG&E:** Author correspondence with PG&E, December 2019.

98 **A gigawatt is a unit:** US Department of Energy estimate, based on National Renewable Energy Labs data showing an average panel size of 320 watts. The solar projects supported by PG&E may have required even more panels, as technology has advanced since then. See Mike Mueller and Mikayla Rumph, "How Much Power Is 1 Gigawatt?," Office of Energy Efficiency & Renewable Energy, August 19, 2019, energy.gov/eere/articles/how-much-power-1-gigawatt.

99 **"The first thing to remember":** Peter Darbee, "Remarks by Peter A. Darbee: Chairman, CEO and President PG&E Corporation; at the United Nations February 14, 2008" (United Nations/Ceres Investor Summit, United Nations, New York, NY, February 14, 2008), transcript, pge-corp.com/investors/investor_info/presentations/20050420_sp.shtml.

99 **The fastest gusts topped eighty-five:** Patricia Grossi, *The 2007 U.S. Wildfire Season: Lessons from Southern California* (Newark, CA: Risk Management Solutions, 2008), 7.

100 **dispatched planes loaded:** County of San Diego, *2007 Firestorms after Action Report*, February 2007, sandiegocounty.gov/oes/docs/2007_SanDiego_Fire_AAR_Main_Document_FINAL.pdf.

100 **One of the pilots:** California Department of Forestry and Fire Protection, *Witch Fire Investigation Report*, October 21, 2007, 19.

100 **Scientists were beginning to recognize:** See, for example, A. L. Westerling, H. G. Hidalgo, D. R. Cayan and T. W. Swetnam, "Warming and Earlier Spring Increase Western U.S. Forest Wildfire Activity," *Science* 313, no. 5789 (August 18, 2006): 940–43, doi.org/10.1126/science.1128834; and A. L. Westerling and B. P. Bryant, "Climate Change and Wildfire in California," in "California at a Crossroads: Climate Change Science Informing Policy," ed. Daniel Cayan, Amy Luers, Guido Franco, Michael Hanemann, Bart Croes, and Edward Vine, supplement, *Climatic Change* 87, no. S1 (March 2008): S231–49, doi.org/10.1007/s10584-007-9363-z.

100 **The idea is to protect:** As quoted in Samir A. Hafez Jr., "Heated Conflict: Investor-Owned Utility Liability for California Wildfires under the Doctrine of Inverse Condemnation," *San Diego Journal of Climate and Energy Law* 11, no. 25

(2020): 25–43, citing Belair v. Riverside County Flood Control Dist., 47 Cal. 3d 550 (1988).

101 **The court determined that private:** Barham v. Southern Cal. Edison Co., 74 Cal. App. 4th 744, 747 (1999).

101 **It would take the CPUC:** CPUC, *Decision Denying Application (A.15-090-10)*, August 22, 2017.

101 **At PG&E's annual shareholder:** Dalzell, interview; and Eric Wolfe, "IBEW Whistleblowers Honored," *Utility Reporter* 57, no. 4 (July/August 2008).

101 **Early on, Accenture had informed:** Accenture, "ESC Benchmarking."

101 **The gas division was suffering:** Jaxon Van Derbeken, "PG&E Incentive System Blamed for Leak Oversights," *San Francisco Chronicle*, December 25, 2011.

102 **The committee agreed:** NorthStar Consulting Group, *Assessment of Pacific Gas and Electric Corporation and Pacific Gas and Electric Company's Safety Culture*, May 8, 2017.

102 **Months later, on the morning:** Report by the National Transportation Safety Board, *Pipeline Accident Brief*, Accident Number: DCA09FP003, adopted May 18, 2010, 1.

103 **The representative called for:** National Transportation Safety Board, *Pipeline Accident Brief*, 2.

103 **A seventeen-year-old girl answered:** California Public Utilities Commission, *Order Instituting Investigation on the Commission's Own Motion into the Operations and Practices of Pacific Gas and Electric Company* (I.10-11-013, Appendix A, section 3.2.1.), November 19, 2010.

103 **In the garage, the grandfather:** CPUC Appendix A.

103 **It registered sixty thousand parts:** CPUC Appendix A.

103 **Moments later, at 1:36 p.m.:** CPUC Appendix A.

103 **Everyone else—daughter, granddaughter:** National Transportation Safety Board, *Pipeline Accident Brief*, 7.

103 **As much as $285 million less than forecast:** Gold, Smith, and Blunt, "Wired to Fail."

103 **Accenture had attempted to centralize:** Unpublished mock-up of *Utility Reporter*, special edition (Summer 2008), reviewed by author.

104 **Even payroll had gotten bungled:** Dalzell, interview.

104 **One summer day in 2008:** Dalzell, interview. Ensuing details from unpublished *Utility Reporter* mock-up.

105 **Accenture charged PG&E more:** Gold, Smith, and Blunt, "Wired to Fail."

105 **"The initiative was too theoretical":** PG&E, *2011 General Rate Case, Prepared Testimony (Ex. PG&E 8, 12A-7)*, December 21, 2009.

105 **Clunky name aside:** Gold, Smith, and Blunt, "Wired to Fail."

105 **If climate change showed Darbee:** Peevey, interview.

105 **It had, however, succeeded:** FactSet analysis by author. Between 2006 and 2010, PG&E's ROE topped 10 percent every year but one.

105 **Darbee later told *The Wall Street Journal*:** Gold, Smith, and Blunt, "Wired to Fail."

106 **The word *transformation*:** Schneider, interview; Dalzell, interview; and Nick Stavropoulos, interview with the author.

Six: San Bruno

107 **A three-thousand-pound piece:** Report by the National Transportation Safety Board, *Pacific Gas and Electric Company Natural Gas Transmission Pipeline Rupture and Fire San Bruno*, NTSB Number: PAR-11-01, adopted August 30, 2011, 16.

107 **The projectile crashed some:** National Transportation Safety Board, *Pacific Gas and Electric Company Natural Gas Transmission Pipeline Rupture*, 16.

107 **The 911 calls poured in:** National Transportation Safety Board, *Pacific Gas and Electric Company Natural Gas Transmission Pipeline Rupture*, 16.

107 **A news helicopter:** See, for example, *CBS Morning News*, September 10, 2010, 3:30 a.m.–4:00 a.m. PST. Accessed through archive.org.

108 **"It's easy to believe it's":** *CBS Morning News*.

108 **The severed pipeline, large:** The pipeline was thirty inches in diameter.

108 **enough to fill more than:** Assuming that the pool is two meters deep, per regulation, an Olympic swimming pool holds 88,286 cubic feet of water.

109 **As sunlight streamed through:** Footage captured on multiple news broadcasts, including *NBC Bay Area News at 5*, September 11, 2010, 4:00 p.m.–4:30 p.m. PST.

109 **"We know our customers":** Lisa M. Krieger, "Upset San Bruno Residents Ask Volleys of Testy Questions of City and Power Company Officials," *San Jose Mercury News*, September 11, 2010.

109 **"Do your people actually have":** Email from Paul Clanon to Brian Cherry and Frank Lindh, "Data on Pipe Segments," October 11, 2010, archived by CPUC as SB_GT&S0013462.

109 **In January, as the investigation:** National Transportation Safety Board, "NTSB Issues Urgent Safety Recommendations as a Result of Preliminary Findings in San Bruno Pipeline Rupture Investigation," news release, January 3, 2011.

110 **Forklifts delivered hundreds:** Associated Press, "Pacific Gas and Electric Using Daly City's Cow Palace to Search through Its Pipeline Documents," *Daily Democrat*, March 6, 2011, dailydemocrat.com/2011/03/06/pgampe-using -cow-palace-to-check-documents/.

110 **As PG&E dusted off:** Nick Stavropoulos, interview with the author.

111 **A crisis occurred:** Stavropoulos, interview.

111 **An eighty-six-year-old distribution:** Report by the Massachusetts Department of Public Utilities, *Incident Report*, report issued June 2007, 2–14, mass .gov/doc/lexington-incident-report/download.

111 **KeySpan's attorneys advised Stavropoulos:** Stavropoulos, interview.

111 **Standing before a group:** Raja Mishra and Dorian Block, "KeySpan Says Worker Error Led to Gas Leaks, Blast," *Boston Globe*, November 11, 2005.

111 **"I take personal responsibility":** Mishra and Block, "KeySpan Says Worker Error Led to Gas Leaks, Blast"; and Stavropoulos, interview.

112 **He wouldn't stand for bullshit:** Stavropoulos, interview.

112 **The company was planning to:** Stavropoulos, interview.

112 **The chairman thought Darbee should:** Russell Gold, Rebecca Smith, and Katherine Blunt, "Wired to Fail," *Wall Street Journal*, December 28, 2019.

112 **He moved in June:** Stavropoulos, interview.

112 **He became a regular:** Stavropoulos, interview.

113 **He arrived alone:** Stavropoulos, interview.

113 **As they put it, PG&E:** Tom Dalzell, interview with the author.

114 **In the union hall, Stavropoulos:** Details, quotes, and photos included in IBEW archives, "Stavropoulos Comes to Weakley Hall," October 10, 2011, ibew 1245.com/2011/10/10/stavropoulos-comes-to-weakley-hall.

115 **As he looked around the:** Stavropoulos, interview.

115 **Tony Earley learned his leadership:** *"Hiring America* featuring Tony Earley," *Hiring America*, November 11, 2013, video, www.youtube.com/watch?v=bJa Jlr7qcHU.

116 **His phone rang:** Daniel Howes, "DTE's Earley Proves Blackout's Bright Spot," *Detroit News*, August 20, 2003.

116 **Earley felt his way down:** Howes, "DTE's Earley Proves Blackout's Bright Spot."

116 **Tie loose and hair:** Mike Hudson and Charles E. Ramirez, "Embattled DTE Tries to Explain," *Detroit News*, August 15, 2003.

116 **"This is an operation":** "Edison: Fixes Could Take Days," Associated Press, August 15, 2003.

116 **Earley worked almost around:** Howes, "DTE's Earley Proves Blackout's Bright Spot."

117 **It expanded the oversight role:** Prior to the passage of the Energy Policy Act of 2005, which established NERC as an electric reliability organization overseen by FERC, NERC's reliability standards were neither mandatory nor enforceable. David Nevius, *The History of the North American Electric Reliability Corporation*, NERC, January 2020, 83, www.nerc.com/AboutNERC/Re source%20Documents/NERCHistoryBook.pdf.

117 **Michigan leaders lauded Earley:** Howes, "DTE's Earley Proves Blackout's Bright Spot."

117 **His instinct was to steer:** Tony Earley, PG&E press conference, December 12, 2011, transcript, 13, pge.com/includes/docs/pdfs/about/newsroom/Press _Conference_12_12_2011.pdf.

118 **The pipe in question:** National Transportation Safety Board, *Pacific Gas and Electric Company Natural Gas Transmission Pipeline Rupture*, 28.

118 **Earley soon announced:** Earley, PG&E press conference, 8.

118 **Earley told investors to expect:** Tony Earley, PG&E earnings call, November 3, 2011, transcript, 4–5.

118 **The CPUC authorized the company:** CPUC, *Proposed Decision of ALJ Bushey* (Rulemaking 11-02-019), February 24, 2011, docs.cpuc.ca.gov/PublishedDocs /Published/G000/M083/K872/83872154.pdf.

118 **"We need to be brutally":** Earley, PG&E press conference, 6–7.

118 **Earley himself was there:** Rachel Maddow and Ezra Klein, *The Rachel Maddow Show*, December 21, 2011, MSNBC.

119 **"We are snakebit":** Brian Cherry to Timothy Alan Simon, "49ers Monday Night Football Readiness Plan," December, 19, 2011, archived by CPUC as SB GT&S 0218256.

119 **"Give Tony my best":** Simon to Cherry, December 19, 2011, archived by CPUC as SB GT&S 0218256.

119 **In October 2011:** John Cox, "PG&E Replaces and Retests Pipeline That Burst Monday," *Bakersfield Californian*, October 28, 2011.

119 **It was one of three:** PG&E, "PG&E Completes 2011 Hydrostatic Pressure Testing Program on Natural Gas Pipelines," news release, December 16, 2011.

119 **In 2012, the company brought:** Stavropoulos, interview.

120 **At that point, only:** Stavropoulos, interview.

120 **He was struck by what:** Stavropoulos, interview.

120 **A ninety-foot wall:** "PG&E Opens State-of-the-Art Gas Control Center in San Ramon," YouTube video, 4:29, posted by "pgevideo," September 12, 2013, youtube.com/watch?v=IpSOTvQoKi8.

120 **He would walk around:** Stavropoulos, interview; and author interviews with former gas division employees.

120 **PG&E traditionally had spent five:** Figures drawn from FERC forms 1 and 2, annual filings detailing maintenance spending in each area.

121 **Lloyd's returned in 2014:** PG&E, "Lloyd's Register Recertifies PG&E as Best-in-Class Gas System Operator," news release, November 12, 2014.

121 **Stavropoulos suggested to colleagues:** Stavropoulos, interview.

Seven: No Free Lunch

123 **The summer of 2008:** "U.S. Drought Summary 2008," National Centers for Environmental Information, National Oceanic and Atmospheric Administration, ncdc.noaa.gov/sotc/drought/200813.

123 **That June, an earthshaking:** California Department of Forestry and Fire Protection, 2008 Wildfire Activity Statistics, fire.ca.gov/media/10885/2008 _wildfireactivitystatistics_complete_revised.pdf.

123 **One never has resources:** Tami Abdollah, "'Critical Day' for Growing Goleta Fire; Big Sur Blaze Only 5% Contained," *Los Angeles Times*, July 5, 2008.

123 **In late 2008, a few:** California Public Utilities Commission, *Order Instituting Rulemaking to Revise and Clarify Commission Regulations Relating to the Safety of Electric Utility and Communications Infrastructure Provider Facilities* (R.08-11-005), November 6, 2008.

124 **Schwarzenegger declared a state:** Bettina Boxhall, "Schwarzenegger Proclaims State of Emergency because of Drought," *Los Angeles Times*, February 28, 2009.

124 **Certain parties, the staffers wrote:** CPUC, *The Consumer Protection and Safety Division's Proposed Rules to Be Implemented in Time for the 2009 Fall Fire Season* (R.08-11-005), 2. Many of the pertinent aspects of this proceeding were very well documented here: Katie Worth, Karen Pinchin, and Lucie Sullivan, "'Deflect,

Delay, Defer': Decade of Pacific Gas & Electric Wildfire Safety Pushback Preceded Disasters," *Frontline* and KQED, August 18, 2020, pbs.org/wgbh/frontline /article/pge-california-wildfire-safety-pushback.

125 **CPUC staffers saw it:** CPUC, *Consumer Protection and Safety Division's Proposed Rules*, 17.

125 **Snowmelt accounts for:** Ali Stevens and Will Chong, "California Drought: 2011–2017," National Integrated Drought Information System, April 26, 2018, noaa.maps.arcgis.com/apps/Cascade/index.html?appid=0307d687789c4d1cb ec397d0abc2fffc.

125 **That water also hydrates forests:** Kelly E. Gleason, John B. Bradford, Anthony W. D'Amato, Shawn Fraver, Brian J. Palik, and Michael A. Battaglia, "Forest Density Intensifies the Importance of Snowpack to Growth in Water-Limited Pine Forests," *Ecological Applications* 31, no. 1 (2021): e02211, doi.org /10.1002/eap.2211.

126 **In a thick filing that:** PG&E, *Opening Comments of Pacific Gas and Electric Company (U39e) on Proposed Decision in Phase 2*, R.08-11-005, 17.

127 **But drafting the legislation proved:** Schwarzenegger to Mary D. Nichols (chairperson, California Air Resources Board), July 15, 2010, arb.ca.gov/regact /2010/res2010/schwarzenegger.pdf.

127 **"Instead of taking oil":** Dana Hull, "Brown Raises Bar for Clean Energy to 33%," *San Jose Mercury News*, April 13, 2011.

128 **office had agreed to:** US Department of Energy, "Energy Department Announces $1.2 Billion Loan Guarantee to Support California Concentrating Solar Power Plant," press release, September 13, 2011, energy.gov/articles /energy-department-announces-12-billion-loan-guarantee-support-california -concentrating.

128 **CPUC staffers crunched some numbers:** CPUC, The Padilla Report to the Legislature (Table A-1: Weighted Average TOD-Adjusted RPS Procurement Expenditures), May 2015.

128 **The staff ultimately recommended either:** CPUC, *Energy Division Resolution E-4433*, October 20, 2011, accessed via docs.cpuc.ca.gov/PUBLISHED /COMMENT_RESOLUTION/143618.htm.

128 **He liked the idea of:** CPUC hearing, November 11, 2011, audio recording, 1:32:20, accessed via cpuc.ca.gov/about-cpuc/transparency-and-reporting /cpuc-voting-meetings/audio-archives.

129 **In the days before:** Mike Florio, interview with the author; and Michael Peevey, interview with the author.

129 **Brown's chief of staff:** Florio, interview.

129 **"Future RPS bidders":** CPUC hearing, 1:34:03.

130 **"I cannot in good conscience":** CPUC hearing, 1:37:20.

130 **exact terms of that contract:** CPUC, *The Padilla Report to the Legislature*, May 2015.

130 **By 2012, PG&E was spending:** Author correspondence with PG&E; and company RPS plans filed with the CPUC.

130 **That spending was projected:** Correspondence with PG&E; and RPS plans.

130 **"There is no free lunch":** California Senate Energy, Utilities and Communications Committee informational hearing, featuring testimony of Michael Peevey, August 7, 2012, video, 34:40, featuring discussion of rate pressure and climate change.

131 **Scientists said, unequivocally:** A. Park Williams, Richard Seager, John T. Abatzoglou, Benjamin I. Cook, Jason E. Smerdon, and Edward R. Cook, "Contribution of Anthropogenic Warming to California Drought during 2012–2014," *Geophysical Research Letters* 42, no. 16 (August 2015): 6819–28.

131 **Governor Jerry Brown declared a:** Office of Edmund G. Brown Jr., "Governor Brown Declares Drought State of Emergency," news release, January 17, 2014, www.ca.gov/archive/gov39/2014/01/17/news18368/index.html.

131 **Tens of millions of coniferous:** USDA Office of Communications, "Forest Service Survey Finds Record 66 Million Dead Trees in Southern Sierra Nevada," news release, June 22, 2016, www.fs.usda.gov/news/releases/forest-service -survey-finds-record-66-million-dead-trees-southern-sierra-nevada. The US Forest Service found that between 2010 and 2015, forty million trees died statewide.

131 **"Unlike Southern California, the need":** CPUC, *Decision Adopting Regulations to Reduce Fire Hazards Associated with Overhead Power Lines and Communication Facilities (D.12-01-032),* January 12, 2012, 49.

132 **The commission instead directed PG&E:** CPUC, *Decision Adopting Regulations to Reduce Fire Hazards Associated with Overhead Power Lines and Communication Facilities,* 3.

132 **In its 2012 decision:** CPUC, *Decision Adopting Regulations to Reduce Fire Hazards,* 50.

132 **It decided that the likelihood:** PG&E to CPUC, "Advice Letter 4167-E," December 21, 2012, included as Attachment B.

132 **The conclusion had several fundamental:** This was a backward-looking analysis. PG&E determined the number of red flag warnings that had occurred in its territory since 2011 and overlaid that data with hourly weather data collected during the same period. It counted more than 9 million wind gusts, 209,911 of which were "concurrent in time and space" with a red flag warning. Thirty-three of them were strong enough to exceed the maximum allowable working stresses for its power lines for a frequency of 0.016 percent. The company concluded that it was "not reasonably foreseeable" that the probability of three-second wind gusts would exceed working stresses defined in General Order 95 by 3 percent or more during a fifty-year period, despite the fact that the infrastructure would continue to wear out over this period. General Order 95 sets rules for overhead electric line construction and can be found here: CPUC, *Overhead Electric Line Construction,* January 2020, docs .cpuc.ca.gov/PublishedDocs/Published/G000/M338/K730/338730245.pdf.

133 **"We all have to admit":** Mike Florio, "Now a PUC Commissioner, Utilities Critic Mike Florio Speaks Out," interview by Jon Brooks, KQED, January 27, 2011, kqed.org/news/14402/new-commissioner-mike-florio-speaks-to-press -after-todays-puc-meeting.

NOTES   325

133 **The consultant issued a report:** Business Advantage Consulting, *California Public Utilities Commission Safety Culture Change Project: Initial Discovery Report*, January 25, 2013.

134 **At the time, the CPUC's:** Katherine Blunt and Russell Gold, "'Safety Is Not a Glamorous Thing': How PG&E Regulators Failed to Stop Wildfire Crisis," *Wall Street Journal*, December 8, 2019.

134 **It mainly focused on auditing:** Blunt and Gold, "'Safety Is Not a Glamorous Thing.'"

134 **The CPUC then created:** Blunt and Gold, "'Safety Is Not a Glamorous Thing.'"

134 **Agency jobs paid less:** Peevey, interview.

134 **Peevey acknowledged that the agency:** Peevey, interview.

134 **"Didn't realize Keenan and Salas":** Michael Peevey to Brian Cherry, April 6, 2011, archived by CPUC as SB_GT&S_0019096.

135 **PG&E depended largely on reliability:** PG&E, *2014 Annual Electric Reliability Report (D.96-09-045, D.04-10-034 and Advice Letter 3812-E)*, February 27, 2015.

135 **For that reason, PG&E often:** PG&E, *2011 General Rate Case (Prepared testimony of Kevin Dasso, 1-40, Exhibit PG&E-3)*, December 21, 2009, 16–32.

135 **Much of it was fast:** PG&E, *Eighteenth Transmission Owner Rate Case Deposition of David Gabbard (FERC Docket No. ER16-2320-002)*, December 4, 2017, 64.

136 **a "pig in the python":** PG&E, *Eighteenth Transmission Owner Rate Case*, 64.

136 **It immediately discovered the same:** Quanta Technology, *Structures (FERC docket ER16-2320-002)*, May 2010.

137 **The CPUC's rules for distribution:** CPUC, *Order Instituting Rulemaking for Electric Distribution Facility Standard Setting (D.12-01-032, amended in R.08-11-00596-11-004)*, January 12, 2012.

137 **The rules for transmission lines:** CPUC, *Order Instituting Rulemaking for Electric Distribution Facility Standard Setting*. The three sentences are as follows: Each utility shall prepare and follow procedures for conducting inspections and maintenance activities for transmission lines. Each utility shall maintain records of inspection and maintenance activities. Commission staff shall be permitted to inspect records and procedures consistent with Public Utilities Code Section 314 (a).

138 **In 2013, PG&E told federal:** PG&E, *Seventeenth Transmission Owner Rate Case (FERC docket no. ER15-2294)*, July 29, 2015, 344–46.

138 **It spelled out the scope:** Paul Marotto to Erika Brenzovich and Mary Sullivan, "NERC program - PG&E Caribou-Big Bend 115kv T/L - Plumas NF Pre-Notification Meeting Proposal." Email obtained through Freedom of Information Act request.

138 **It would be delayed again:** Author review of PG&E transmission owner cases filed with FERC between 2012 and 2019. See Katherine Blunt and Russell Gold, "PG&E Delayed Safety Work on Power Line That Is Prime Suspect in California Wildfire," *Wall Street Journal*, February 28, 2019.

138 **It was one of many:** PG&E transmission owner cases. See Katherine Blunt and Russell Gold, "PG&E Knew for Years Its Lines Could Spark Wildfires, and Didn't Fix Them," *Wall Street Journal*, July 10, 2019.

138 **Meanwhile, the CPUC launched:** CPUC, *Order Instituting Rulemaking to Develop and Adopt Fire-Threat Maps and Fire-Safety Regulations*, R-15-05-006, May 7, 2015.

139 **"No matter the menu":** Brian Cherry to Michael Peevey, May 29, 2010, archived by the CPUC as SB GT&S 0003248.

139 **"This is a major problem":** Brian Cherry to Carol Brown, January 17, 2014, archived by the CPUC as SB GT&S 0328240.

139 **Peevey never saw it that way:** Peevey, interview.

140 **The idea of settlements:** San Diego Tribune Editorial Board, "Meet Michael Picker, President of the Beleaguered California Public Utilities Commission," *San Diego Tribune*, September 7, 2018.

140 **"If there's a prohibition":** CPUC Voting Meeting, January 15, 2015, video, 12:00, featuring Picker's introductory remarks as president, www.adminmonitor.com /ca/cpuc/voting_meeting/20150115/.

140 **He called the incriminating emails:** CPUC Voting Meeting, 7:43.

141 **That year, the agency required:** Fire incident data reported to CPUC, uploaded as "Pacific Gas and Electric Company Fire Incident Data Collection Plan," accessed through cpuc.ca.gov/-/media/cpuc-website/files/uploadedfiles /cpucwebsite/content/news_room/newsupdates/2020/pge-fire-incident-data -2014-2019.pdf.

141 **PG&E wasn't keen:** Fire incident data reported to CPUC.

Eight: The Trial

146 **One interview was especially revealing:** Gary Harpster, interview by James Haggarty, San Bruno Police Department, San Bruno, CA, September 15, 2012.

147 **A couple weeks later, Haggarty:** Frank Maxwell, interview by James Haggarty, San Bruno Police Department, San Bruno, CA, September 27, 2012.

147 **PG&E then decided the division:** These are approximate figures. In one email, Frank Maxwell wrote that the gas transmission division was expected to receive $95 million of the $120 million request it submitted to the CPUC. Maxwell wrote in another email that the gas transmission the CPUC authorized 94.7 percent of the amount authorized by the CPUC. The CPUC authorized about $95 million, so it can be deduced that PG&E allowed the division about $90 million.

147 **In late 2007, the manager:** Frank Maxwell to Chris Warner, Thomas Robinson, Robert Fassett, Glen Carter, Les Buchner, and Pam Johnson, "Urgent: Minimum Funding Levels for IM 2008-2010," November 2, 2007, Bates stamp PGE_DOJ_0264036.

148 **The company responded to subpoena:** Transcript of Proceedings, Document 8, 4, April 23, 2014, 3:14-cr-00175-WHA.

148 **In April 2014, prosecutors charged:** Indictment, Doc. 1, April 1, 2014, *United States of America vs. Pacific Gas & Electric Company*, 3:14-cr-00175-WHA.

149 **On April 21, 2014, PG&E:** Transcript of Proceedings, Document 12, April 24, 2014, 3:14-cr-00175-WHA.

149 **Tony Earley, the company's CEO:** Earley, PG&E first-quarter earnings call, May 1, 2014, transcript, S&P Global Market Intelligence.

150 **If convicted, the company wouldn't:** PG&E reported about $1.44 billion in net income in 2014. PG&E, Annual Report Pursuant to Section 13 or 15(d) of the Securities Exchange Act of 1934 (Form 10-K), February 10, 2015, www.sec .gov/Archives/edgar/data/0001004980/000100498015000010/form10k.htm.

150 **The grand jury, after reconvening:** James Haggarty, interview with the author.

150 **In July, the prosecutors filed:** Superseding indictment, Document 22, July 30, 2014, 3:14-cr-00175-WHA.

151 **"Did people make bad judgments?":** Bob Egelko, "PG&E Pleads Not Guilty in San Bruno Blast Case," *San Francisco Chronicle*, August 18, 2014.

151 **"Corporations have neither bodies":** Thurlow is sometimes quoted as having said, "Did you ever expect a corporation to have a conscience, when it has no soul to be damned, and no body to be kicked?" John C. Coffee Jr., 'No Soul to Damn: No Body to Kick': An Unscandalized Inquiry into the Problem of Corporate Punishment," *Michigan Law Review* 79, no. 3 (January 1981): 386–459.

152 **corporations have the constitutional right:** Bank of United States v. Deveaux, 9 U.S. (5 Cranch) 61 (1809). The significance of this precedent is masterfully explained in Adam Winkler, *We the Corporations: How American Businesses Won Their Civil Rights* (New York: Liveright, 2018), 76–77.

152 **It remains an evolving:** Citizens United v. Federal Election Comm'n, 558 U.S. 310 (2010). The court ruled that the government could not suppress political speech on the basis of the speaker's identity as a corporation; Burwell v. Hobby Lobby Stores, Inc., 134 S. Ct. 2751 (2014). The court ruled that owners of closely held corporations with sincere religious beliefs about contraception have the right to deny their employees health insurance coverage for contraception.

152 **The idea of corporate liability:** New York C. & H. R. R. Co. v. United States, 212 U.S. 481, 496 (1909).

152 **In the court's view, corporations:** Opinion by Justice William R. Day, accessed at caselaw.findlaw.com/us-supreme-court/212/481.html.

152 **In the 1990s, a federal:** Christopher A. Wray, "Corporate Probation under the New Organizational Sentencing Guidelines," *Yale Law Journal* 101, no. 8 (2017): 2018–42.

154 **company had agreed to pay:** Ari Bloomekatz, "PG&E to Pay $565 Million to Victims in San Bruno Pipeline Explosion," *Los Angeles Times*, September 10, 2013.

154 **Picker was unsympathetic:** CPUC, *President Proposes $1.6 Billion in Remedies in PG&E Pipeline Cases, Increasing Penalty Amount by $200 Million and Directing $850 Million to Shareholder-Funded Safety Improvements (Docket I.12-01-007)*, March 13, 2015.

155 **The question was whether employees:** The significance of this case is brilliantly explained by Brandon L. Garrett in *Too Big to Jail: How Prosecutors Compromise with Corporations* (Cambridge, MA: Harvard University Press, 2014), 19–44.

155 **In their words:** Larry Thompson, deputy attorney general news conference, Arthur Andersen indictment, March 14, 2002, Department of Justice Conference Center, transcript, justice.gov/archive/dag/speeches/2002/031402news conferncearthurandersen.htm.

155 **Their black shirts were stamped:** James Nielsen, *Employees of Arthur Andersen Lead Protest across the Street*, May 20, 2002, photograph, accessed via Getty Images, gettyimages.com/detail/news-photo/employees-of-arthur-andersen-lead -protest-across-the-street-news-photo/51703555?adppopup=true.

156 **Arthur Andersen's lead defense counsel:** Alexei Barrionuevo and Jonathan Weil, "Andersen Attorney Steals the Show, but Will Hardin's Antics Pay Off?," *Wall Street Journal*, May 20, 2002; photo by James Nielsen/AFP via Getty Images, June 8, 2002.

156 **"You know the little routine":** As quoted in Garrett, *Too Big to Jail*, 28.

156 **"Are you Waldo?":** Barrionuevo and Weil, "Andersen Attorney Steals the Show."

156 **He sounded a warning:** Garrett, *Too Big to Jail*, 29.

156 **The firm had severed ties:** Garrett, *Too Big to Jail*, 40.

158 **The activist was Martin Luther King Jr.:** Abbie Vansickle, "The Catalyst: Thelton Henderson Transformed California's Criminal Justice System. Now Comes the Backlash," *The Marshall Project*, April 23, 2018.

159 **the statute of limitations:** Mot. to Dismiss Counts 4, 5 and 24-28 as Barred by the Statute of Limitations, Doc. 113, July 27, 2015, *United States of America vs. Pacific Gas & Electric Company*, 3:14-cr-00175-WHA.

159 **Next, it sought to dismiss:** Mot. to Dismiss for Multiplicity, Doc. 124, September 7, 2015, *United States of America vs. Pacific Gas & Electric Company*, 3:14 -cr-00175-WHA.

159 **Even if the company were:** Mot. to Dismiss for Multiplicity, Doc. 124.

160 **By that logic, whether PG&E:** Mot. to Dismiss for Failure to State an Offense, Doc. 125, September 25, 2015, *United States of America vs. Pacific Gas & Electric Company*, 3:14-cr-00175-WHA.

160 **Precedent wasn't clear:** Mot. to Dismiss for Erroneous Legal Instructions to the Grand Jury, Doc. 127, 8-12, September 7, 2015, *United States of America vs. Pacific Gas & Electric Company*, 3:14-cr-00175-WHA.

161 **That warm afternoon, the spindly:** Cal Fire, *Investigation Report (Butte Incident)*, April 25, 2016, 16–17.

161 **It wasn't the Diablo winds:** Winds blew at four to five miles per hour, per *Investigation Report (Butte Incident)*, 11.

161 **PG&E contractors had come by:** *Investigation Report (Butte Incident)*, 4.

161 **The company's equipment sparked:** Fire incident data reported to CPUC, uploaded as "Pacific Gas and Electric Company Fire Incident Data Collection Plan," accessed through cpuc.ca.gov/-/media/cpuc-website/files/uploadedfiles /cpucwebsite/content/news_room/newsupdates/2020/pge-fire-incident-data -2014-2019.pdf.

162 **He had agreed to tape:** Greg Dalton, "Charging Ahead: PG&E CEO Tony Earley," October 15, 2015, in *Climate One*, 59:06, climateone.org/audio/charging -ahead-pge-tony-earley.

162 **PG&E was paying an average:** CPUC, *The Padilla Report to the Legislature, (Table A-2.Weighted Average TOD-Adjusted RPS Procurement Expenditures)* 2015.

163 **The defense notched a:** Order by the Hon. Thelton E. Henderson, Doc. 217, December 23, 2015, *United States of America vs. Pacific Gas & Electric Company*, 3:14-cr-00175-WHA.

164 **Twelve people lined up:** Volume 1, Transcript of Trial Proceedings, Doc. 667, 51, June 16, 2016, *United States of America vs. Pacific Gas & Electric Company*, 3:14-cr-00175-WHA.

164 **Those remaining:** Volume 1, Transcript of Trial Proceedings, Doc. 667, 76, June 16, 2016, *United States of America vs. Pacific Gas & Electric Company*, 3:14-cr-00175-WHA.

164 **The defense team told Henderson:** Mot. Regarding Jury Selection, Doc. 641, 1, June 12, 2016, *United States of America vs. Pacific Gas & Electric Company*, 3:14-cr-00175-WHA.

164 **It pushed for a questionnaire:** Mot. Defendant's Request for Jury Questionnaire and Certain Other Jury Selection Procedures, Doc. 297, February 22, 2016, *United States of America vs. Pacific Gas & Electric Company*, 3:14-cr-00175-WHA.

164 **filled out sixteen pages:** Mot. Defendant's Request for Jury Questionnaire and Certain Other Jury Selection Procedures, Doc. 297, Attachment 1, February 22, 2016, *United States of America vs. Pacific Gas & Electric Company*, 3:14-cr-00175-WHA.

165 **One of the first to:** Volume 1, Transcript of Trial Proceedings, Doc. 667, 113, June 16, 2016, *United States of America vs. Pacific Gas & Electric Company*, 3:14-cr-00175-WHA.

165 **The next prospective juror:** Volume 1, Transcript of Trial Proceedings, Doc. 667, 126, June 16, 2016, *United States of America vs. Pacific Gas & Electric Company*, 3:14-cr-00175-WHA.

166 **Then he shifted his line:** Volume 2, Transcript of Trial Proceedings, Doc. 668, 301, June 16, 2016, *United States of America vs. Pacific Gas & Electric Company*, 3:14-cr-00175-WHA.

166 **Bauer was ready to weed:** Volume 2, Transcript of Trial Proceedings, Doc. 668, 301, June 16, 2016, *United States of America vs. Pacific Gas & Electric Company*, 3:14-cr-00175-WHA.

166 **"Due to *Citizens United*":** Volume 2, Transcript of Trial Proceedings, Doc. 668, 302, June 16, 2016, *United States of America vs. Pacific Gas & Electric Company*, 3:14-cr-00175-WHA.

166 **"I do feel that some":** Volume 2, Transcript of Trial Proceedings, Doc. 668, 302, June 16, 2016, *United States of America vs. Pacific Gas & Electric Company*, 3:14-cr-00175-WHA.

167 **"That's pretty much a trend":** Volume 2, Transcript of Trial Proceedings, Doc. 668, 338, June 16, 2016, *United States of America vs. Pacific Gas & Electric Company*, 3:14-cr-00175-WHA.

167 **Judge Henderson gave them:** Volume 3, Transcript of Trial Proceedings, Doc. 669, 485, June 16, 2016, *United States of America vs. Pacific Gas & Electric Company*, 3:14-cr-00175-WHA.

168 **On a screen:** Transcript of Proceedings as to Pacific Gas and Electric Company, Vol. 27, Doc. 838, 101, July 26, 2016, *United States of America vs. Pacific Gas & Electric Company*, 3:14-cr-00175-WHA.

172 **West then displayed a PowerPoint:** Transcript of Proceedings, July 7, 2016, Volume 16, Doc. 731, 2338, *United States of America vs. Pacific Gas & Electric Company*, 3:14-cr-00175-WHA.

173 **At issue was a memo:** Transcript of Proceedings, held on July 8, 2016, Volume 17, Doc. 745, 2491, *United States of America vs. Pacific Gas & Electric Company*, 3:14-cr-00175-WHA.

173 **But it hadn't done so:** Transcript of Proceedings, held on July 8, 2016, Volume 17, Doc. 745, 2499, *United States of America vs. Pacific Gas & Electric Company*, 3:14-cr-00175-WHA.

173 **So Manegold, after consulting with:** Transcript of Proceeding, Volume 17, Doc. 745, 2491; 0571, July 8, 2016, *United States of America vs. Pacific Gas & Electric Company*, 3:14-cr-00175-WHA.

173 **Although federal regulations state:** Transcript of Proceeding, Volume 17, Doc. 745, 2494; Ex. 0571, July 8, 2016, *United States of America vs. Pacific Gas & Electric Company*, 3:14-cr-00175-WHA.

173 **Ahead of the audit, he:** Transcript of Proceeding, Volume 17, Doc. 745, 2502; Ex. 0575, July 8, 2016, 3:14-cr-00175-WHA.

174 **"Bill, sounds reasonable":** Transcript of Proceeding, Volume 17, Doc. 745, 2506; Ex. 0578, July 8, 2016, *United States of America vs. Pacific Gas & Electric Company*, 3:14-cr-00175-WHA.

174 **Manegold responded in jest:** Transcript of Proceeding, Volume 17, Doc. 745, 2506; Ex. 0578, July 8, 2016, 3:14-cr-00175-WHA.

174 **When he returned to his:** Transcript of Proceeding, Volume 17, Doc. 745, 2505, July 8, 2016, 3:14-cr-00175-WHA.

174 **By his own admission, he:** Transcript of Proceeding, Volume 17, Doc. 745, 2505, July 8, 2016, *United States of America vs. Pacific Gas & Electric Company*, 3:14-cr-00175-WHA. (Question to Manegold: Did you feel comfortable with the language that you had written in that white paper? Response: No. Question: Did you think PG&E should stand by it? Response: No.)

174 **Manegold exchanged emails:** Transcript of Proceeding, Volume 17, Doc. 745, 2520-26, July 8, 2016, *United States of America vs. Pacific Gas & Electric Company*, 3:14-cr-00175-WHA.

175 **His friend answered:** Transcript of Proceeding, Vol. 17, Doc. 745, 2524, July 8, 2016, *United States of America vs. Pacific Gas & Electric Company*, 3:14-cr-00175-WHA.

176 **They were ready to be:** Transcript of Proceeding, Vol. 20, Doc. 753, 2822, July 13, 2016, *United States of America vs. Pacific Gas & Electric Company*, 3:14-cr-00175-WHA.

177 **That day, Henderson instructed:** Order Regarding Jury Instructions on Intent Elements for Regulatory Counts, Doc. 826, July 26, 2016, *United States of America vs. Pacific Gas & Electric Company*, 3:14-cr-00175-WHA.

178 **Hoffman didn't think:** Tom Hoffman, interview with the author.

179 **Haggarty felt a sort:** Haggarty, interview.

180 **In January, Henderson issued:** Judgment in a Criminal Case as to Pacific Gas and Electric Company, Doc. 992, July 26, 2016, *United States of America vs. Pacific Gas & Electric Company*, 3:14-cr-00175-WHA.

Nine: Fire Siege

182 **Why not her?:** Todd Johnson and Emily Fancher, "To Advance, PG&E's New Chief Asked 'Why Not Me?,'" *San Francisco Business Times*, May 4, 2017.

183 **A California Public Utilities Commission staffer:** Michael Campbell to Sidney Dietz, July 2, 2014, archived by CPUC as SB GT&S 0272487.

183 **On an earnings call:** Geisha Williams, PG&E Corp. first-quarter earnings call, May 2, 2017, transcript, S&P Global Market Intelligence.

183 **That involved laying off eight:** PG&E, "PG&E Streamlining Management, Implementing Efficiency Measures to Keep Customer Bills Affordable while Investing in the Future," press release, January 11, 2017.

183 **"You can expect to see more":** Williams, PG&E Corp. first-quarter earnings call.

184 **186 fires had started:** Charles Filmer, *Summary and Analysis of VM Fire Incidents (2007-2010)*, January 2011. Confidential document reviewed by author.

184 **In a presentation that year:** PG&E, "Electric Operations Enterprise Wildfire Risk Update" (presentation, PG&E board of directors meeting, December 16, 2014). Confidential presentation reviewed by author.

185 **It resorted to allowing one:** Deposition of Brian Joiner (president of ACRT Pacific), Volume I, 40–53, May 17, 2017 (Ca.Super. Ct. JCCP 4853).

185 **That same contractor:** Deposition of Pete Dominguez (PG&E vegetation management specialist) Volume 1, 136; Volume 3, 234, March 31, 2017 (Ca.Super. Ct. JCCP 4853); and deposition of Bob Bell (former manager of vegetation management, PG&E), Volume 1, 30-45; 70-71, June 2, 2017 (Ca.Super. Ct. JCCP 4853).

185 **In 2014, PG&E's head:** Deposition of Eric Back (former vegetation management director for PG&E), Volume 1, 117, April 27, 2017 (Ca.Super. Ct. JCCP 4853).

186 **Will Abrams awoke:** Will Abrams, interview with the author, in which he described the ensuing details of his family's escape.

188 **The call would be different:** Geisha Williams, PG&E Third-Quarter earnings call, November 2, 2017, transcript, S&P Global Market Intelligence.

188 **At the end of November:** CPUC, *Decision Denying Application (D.18-07-025)*, November 30, 2017, docs.cpuc.ca.gov/PublishedDocs/Published/G000/M218/K019/218019946.pdf.

189 **On December 20, the company:** PG&E, "PG&E Announces Suspension of Dividend, Citing Uncertainty Related to Causes and Potential Liabilities Associated with Northern California Wildfires," press release, December 20, 2017, investor.pgecorp.com/news-events/press-releases/press-release-details/2017/PGE-Announces-Suspension-of-Dividend-Citing-Uncertainty-Related-to

-Causes-and-Potential-Liabilities-Associated-with-Northern-California
-Wildfires/default.aspx.

190 **Initially, he didn't ask much:** Minute Entry, Doc. 940, August 15, 2017, *United States of America vs. Pacific Gas & Electric Company*, 3:14-cr-00175-WHA.

190 **In court, he addressed:** Transcript of Proceedings, Doc. 947, 9, December 7, 2017, 3:14-cr-00175-WHA.

190 **"Your Honor," the monitor replied:** Transcript of Proceedings, Doc. 947, 10, December 7, 2017, 3:14-cr-00175-WHA.

191 **PG&E swiftly came out:** PG&E, "PG&E Statement on California Senate Bill 819," press release, January 3, 2018, pgecurrents.com/2018/01/03/pge-statement -on-california-senate-bill-819.

191 **The fires, she said:** PG&E, "PG&E CEO Geisha Williams Highlights California's Clean Energy Progress and Key Challenge for the Future," press release, January 31, 2018.

191 **The final product:** CPUC Fire-Threat Map, adopted January 19, 2018, accessed via cafirefoundation.org/cms/assets/uploads/2020/05/CPUC_Fire-Threat_Map _final.pdf.

192 **Meanwhile, PG&E had been reporting:** Fire incident data reported to CPUC, uploaded as "Pacific Gas and Electric Company Fire Incident Data Collection Plan," accessed through cpuc.ca.gov/-/media/cpuc-website/files/up loadedfiles/cpucwebsite/content/news_room/newsupdates/2020/pge-fire -incident-data-2014-2019.pdf.

192 **In March, PG&E announced:** PG&E, "In Advance of 2018 Wildfire Season, PG&E Takes Action with Comprehensive Community Wildfire Safety Program," press release, March 22, 2018.

192 **released some of the results:** PG&E SEC Form 8-K, filed June 8, 2018.

192 **an emergency notice to investors:** PG&E SEC Form 8-K, filed June 21, 2018.

192 **Analysts dialed in:** PG&E Corp. special call, June 21, 2018.

193 **Williams sat for hers in July:** Videotaped Deposition of Geisha Jimenez Williams (Transcript), 36–38, July 7, 2017 (Ca.Super. Ct. JCCP 4853).

193 **Around the same time, Nick Stavropoulos:** PG&E, "PG&E's Stavropoulos to Retire at the End of Third Quarter 2018," press release, July 13, 2018, https:// pgecurrents.com/2018/07/13/pges-stavropoulos-to-retire-at-the-end-of-third -quarter-2018/

193 **But he had come to:** Nick Stavropoulos, interview with the author.

193 **After the 2017 fires, the:** CPUC, *Decision Adopting Regulations to Enhance Fire Safety in the High-Threat Fire District (D 17-12-024)*, December 14, 2017, https:// docs.cpuc.ca.gov/PublishedDocs/Published/G000/M200/K976/200976667.pdf.

194 **One of the most significant measures:** Senate Bill No. 901, Ch. 626, Statutes of 2018 (2018).

194 **PG&E estimated that an average:** Erin Ailworth and Sara Randazzo, "California Passes Bill to Rescue Utility Facing Fire Costs," *Wall Street Journal*, September 1, 2018.

195 **Its wildfire operations center:** PG&E, *Public Safety Power Shutoff Report to the CPUC (events October 13–17, 2018)*, October 31, 2018, 2, https://pge.com

/pge_global/common/pdfs/safety/emergency-preparedness/natural-disaster
/wildfires/PSPS-Report-Letter-11.27.18.pdf.

195 **Calls went out to ninety-seven thousand:** PG&E, *Public Safety Power Shutoff*, 2.

195 **At dinnertime on Sunday:** PG&E, *Public Safety Power Shutoff*, 2. The lights went
out at 8:00 p.m. in the North Bay and in the Sierra foothills later that evening.

195 **The Calistoga Inn, an eighteen-room:** Russell Gold and Katherine Blunt,
"PG&E's Radical Plan to Prevent Wildfires: Shut Down the Power Grid," *Wall
Street Journal*, April 27, 2019.

195 **At the Calistoga Roastery:** Gold and Blunt, "PG&E's Radical Plan to Prevent
Wildfires."

196 **Cal Mart, the local grocery:** Gold and Blunt, "PG&E's Radical Plan to Prevent Wildfires."

196 **All told, PG&E ended up:** PG&E, Public Safety *Public Safety Power Shutoff*, 2.

197 **Hours after the fire started:** PG&E, Electric Safety Incident Reported: Pacific Gas & Electric Incident No: 181108-9002, November 8, 2018.

197 **their steepest plunge:** Katherine Blunt and Russell Gold, "California's Largest
Utility Pummeled by Wildfire Risks," *Wall Street Journal*, November 14, 2018.

197 **exhausted its lines of credit:** PG&E, SEC Form 8-K, November 9, 2018.

198 **At the end of November, Alsup:** Notice Re California Wildfires, Document
951, November 27, 2019, *United States of America vs. Pacific Gas & Electric Company*, 3:14-cr-00175-WHA.

198 **Its attorneys stated the obvious:** Response by Pacific Gas and Electric Company, Document 956, 2, December 31, 2018, *United States of America vs. Pacific
Gas & Electric Company*, 3:14-cr-00175-WHA.

198 **Hallie Hoffman and Jeff Schenk:** Response by USA, Document 955, December 31, 2018, *United States of America vs. Pacific Gas & Electric Company*, 3:14-
cr-00175-WHA.

Ten: Chapter 22

200 **On January 14, 2019, PG&E:** PG&E, Form 8-K, Securities, January 14, 2019,
https://sec.gov/Archives/edgar/data/0001004980/000095015719000032
/form8k.htm.

201 **Already, PG&E faced dozens:** PG&E Form 8-K, January 14, 2019.

202 **more than $30 billion:** PG&E Form 8-K, January 14, 2019.

203 **So, a couple of weeks:** Order to Show Cause, Doc. 961, January 10, 2019,
United States of America vs. Pacific Gas & Electric Company, 3:14-cr-00175-WHA.

204 **PG&E's attorneys filed a lengthy:** Response to Order to Show Cause, Doc.
976, January 23, 2019, *United States of America vs. Pacific Gas & Electric Company*,
3:14-cr-00175-WHA.

205 **The first thing he did:** Author attended this hearing. The back-and-forth
between Alsup and the attorneys is transcribed in Transcript of Proceedings,
Doc. 999, January 31, 2019, *United States of America vs. Pacific Gas & Electric Company*, 3:14-cr-00175-WHA.

205 **He was almost exactly right:** FactSet analysis done by author.

207 **Between 2014 and 2017, as:** Russell Gold, Katherine Blunt, and Rebecca Smith, "PG&E Sparked at Least 1,500 California Fires. Now the Utility Faces Collapse," *Wall Street Journal*, January 13, 2019.

207 **Twenty-four people died:** Two people died in the 2015 Butte Fire, and twenty-two died in the 2017 fires sparked by PG&E power lines.

207 **At the start of February:** PG&E, Amended 2019 Wildfire Safety Plan (R.-18-10-007), Feb. 6, 2019.

208 **A few weeks later, PG&E:** PG&E, Form 10-K, February 28, 2019; and Katherine Blunt and Russell Gold, "PG&E Says Its Equipment Was Probable 'Ignition Point' of Camp Fire, Takes $11.5 Billion in Charges," *Wall Street Journal*, February 28, 2019.

209 **And it had acquired something:** "First Amended Verified Statement of the Ad Hoc Group of Subrogation Claim Holders," Doc. 1482, Ex. A at 3, In re PG&E Corporation, U.S. Bankruptcy Court, Northern District of California, Case No. 19-30088.

209 **A number of companies sold:** Peg Brickley, "Klarman's Baupost Poised to Cash In On PG&E Insurance Bet," *Wall Street Journal*, September 13, 2019.

210 **The proposal was largely unprecedented:** Katherine Blunt, "A Judge Wants to Control PG&E's Dividends until It Reduces Risk of Fires," *Wall Street Journal*, March 31, 2019.

210 **In their response to the:** Response by Pacific Gas & Electric, Document 1032, *United States of America vs. Pacific Gas & Electric Company*, 3:14-cr-00175-WHA.

211 **He marveled at the destruction:** Transcript of Proceedings, Document 1047, 5, April 4, 2019, *United States of America vs. Pacific Gas & Electric Company*, 3:14-cr-00175-WHA.

211 **"It's going to be your":** Transcript of Proceedings, Doc. 1047, 9, April 4, 2019.

212 **By his own assessment:** Bill Johnson, "TVA's Bill Johnson on Leading in the First 100 Days," interview by David Gee, Boston Consulting Group, July 12, 2013, bcg.com/publications/2013/energy-environment-people-organization-tva-bill-johnson-leading-first-100-days.

212 **The jarring experience forced him:** Ed Marcum, "CEO Bill Johnson Calls on his Past to Guide TVA's Future," *Knoxville News Sentinel*, August 31, 2013. Johnson is quoted as follows: "What did I learn from the experience? You should be prepared for whatever life offers you. . . . Your ability to look forward, your personal toughness, your resolve—these are things that will be tested. It is very important when you have an event in life to know who you are and be comfortable with who you are, what you do and how you do it."

212 **In March 2019, Johnson left:** A four-bedroom, three-bathroom home.

213 **At 8:05 a.m. on March:** PG&E, *Notice of Ex Parte Communication (CPUC I-15-08-019)*, March 12, 2019.

213 **He said the company should:** Michael Picker, *Notice of Ex Parte Communication (CPUC I-15-08-019)*, March 12, 2019.

213 **Governor Newsom was also deeply:** Russell Gold, "California Governor Disapproves of Proposed PG&E Board," *Wall Street Journal*, March 28, 2019.

213 **he sent a strongly worded:** Gavin Newsom to John Simon, March 28, 2019, gov.ca.gov/2019/03/28/pge-board.

214 **Protesters gathered outside company headquarters:** Elena Shao, "PG&E Shareholders Approve New Board amid Protests Outside," *San Francisco Chronicle*, June 21, 2019.

214 **They planted flowers:** Lyanne Melendez, "Inside PG&E's Annual Shareholders Meeting," ABC 7 News, June 21, 2019, abc7news.com/pge-new-board-trust /5357748.

214 **Nothing was off the table:** CPUC workshop, November 15, 2018, video, 1:00:00, adminmonitor.com/ca/cpuc/workshop/20181115/.

215 **The meeting opened:** CPUC, "Public Forum on the Future of PG&E," video, 10:00, April 15, 2019, adminmonitor.com/ca/cpuc/workshop/20190415/.

217 **He gave his performance:** Michael Picker, interview with the author.

217 **His task was to pick:** J. D. Morris, "California Public Utilities Commission President Michael Picker to Step Down," *San Francisco Chronicle*, May 30, 2019.

217 **They came up with a:** Alejandro Lazo and Katherine Blunt, "California Legislature Approves Multibillion-Dollar Wildfire Fund," *Wall Street Journal*, July 11, 2019.

218 **In June, it had agreed:** Katherine Blunt and Erin Ailworth, "PG&E Reaches $1 Billion Settlement with Paradise, California Governments," *Wall Street Journal*, June 18, 2019.

218 **It was a daylong affair:** Hearing Transcript, Doc. 3563, August 14, 2019, United States Bankruptcy Court, Northern District of California, Case No. 19-30088.

218 **He had also taken:** Statement of Cravath, Swaine & Moore LLP, Doc. 3484, August 8, 2019, United States Bankruptcy Court, Northern District of California, San Francisco Division, Case No. 19-30088.

219 **It was a crushing feeling:** Will Abrams, interview with the author.

219 **Some days brought vertigo:** Abrams, interview.

219 **Every week brought:** Abrams, interview.

219 **In August, Abrams sent:** Letter to the Court, Doc. 3828, September 6, 2019, United States Bankruptcy Court, Northern District of California, Case No. 19-30088.

220 **But asserting himself:** Abrams, interview.

220 **He wanted to testify personally:** Letter to the Court, Doc. 3906, September 17, 2019, United States Bankruptcy Court, Northern District of California, Case No. 19-30088.

220 **The very next day, PG&E:** Peg Brickley and Russell Gold, "PG&E Reaches $11 Billion Settlement with Insurers over Deadly Wildfires," *Wall Street Journal*, September 13, 2019.

220 **Baupost Group, the hedge fund:** Brickley, "Klarman's Baupost Poised to Cash In."

221 **It was, at that point:** Fourth Amended Statement of the Ad Hoc Group of Subrogation Claim Holders, Document 4228, Ex. A at 4, October 16, 2019, United States Bankruptcy Court, Northern District of California, Case No. 19-30088.

Eleven: Blackout

222 **The room echoed with applause:** CPUC meeting, video, 1:00:00 August 15, 2018, adminmonitor.com/ca/cpuc/voting_meeting/20190815/.

223 **By mid-September, it had worked:** Katherine Blunt and Rebecca Smith, "PG&E's Big Blackout Is Only the Beginning," *Wall Street Journal*, October 12, 2019; and Document 1099, 3, *United States of America vs. Pacific Gas & Electric Company*, 3:14-cr-00175-WHA.

223 **On October 4, 2019, the meteorologists:** PG&E, *Public Safety Power Shutoff (PSPS) Report*, October 25, 2019.

224 **PG&E held an urgent conference:** PG&E, *Public Safety Power Shutoff (PSPS) Report*, 4.

225 **Hundreds of thousands of them:** PG&E, *Public Safety Power Shutoff (PSPS) Report*, Table 1-2.

225 **30,000 customers:** PG&E identified about 30,300 medical baseline customers within the scope of this shutoff.

227 **More than four hundred schools:** Zusha Elinson, Ian Lovett, Alejandro Lazo, and Jim Carlton, "'I'm Out': PG&E Blackouts Stagger Californians," *Wall Street Journal*, October 13, 2019.

229 **A week after the blackouts:** California Public Utilities Commission voting meeting, October 18, 2019, video, adminmonitor.com/ca/cpuc/voting_meeting /20191018.

229 **"Let me assure you":** CPUC voting meeting at 18:30.

231 **The company, after assessing:** PG&E, *Public Safety Power Shutoff (PSPS) Report to the CPUC October 9–12, 2019 De-Energization Event*, Appendix B.

231 **a handful of the old 115-kilovolt lines:** PG&E, *Public Safety Power Shutoff (PSPS) Report to the CPUC*.

231 **On the night of October 23:** Rong-Gong Lin II, Jaclyn Cosgrove, and Paul Duginski, "Dangerous Diablo Winds Fueling Kincade Fire with Gusts up to 76 mph," *Los Angeles Times*, October 24, 2019.

231 **"Sometimes things just break":** "Evidentiary Hearings in PG&E's Bankruptcy Proceeding (I.19-09-016)," California Public Utilities Commission, Admin Monitor video of proceeding (part 2), 52:30-2:53:52, adminmonitor .com/ca/cpuc/hearing/20200225/2.

232 **among the oldest in the system:** Quanta Technology, "Structures," prepared for Pacific Gas & Electric Co., May 2010, filed in TO18 rate case, FERC docket ER16-2320-002.

232 **The company issued a notice:** Posting Requirements for FERC Standards of Conduct, Pacific Gas and Electric Company, October 25, 2019, pge.com/in cludes/docs/pdfs/mybusiness/customerservice/nonpgeutility/electrictrans mission/ownersforum/ferc3587c102519.pdf.

232 **The next day, the lights:** A total of 967,700 customers were affected by these outages. Three million is not a perfect estimate, but it is based on census data showing that the average California household is home to 2.95 people.

233 **Across the state, residents:** Taryn Luna, "Californians Want to End PG&E's Operations as They Exist Now, New Poll Says," *Los Angeles Times*, December 10, 2019.

235 **In bankruptcy court, it submitted:** Katherine Blunt, "San Francisco Offers to Buy PG&E Wires in City," *Wall Street Journal*, September 8, 2019.

235 **On October 7, as PG&E:** William Johnson to London Breed and Dennis Herrera, "Re: City and County of San Francisco Offer to Acquire Certain PG&E Retail Electric Facilities," October 7, 2019.

236 **Watched the fallout with sadness:** Dan Richard, interview with the author.

236 **PG&E had become his white whale:** Richard, interview.

237 **still more than eight hundred:** Estimate by the National Rural Electric Cooperative Association, a trade group.

238 **In a pitch explaining:** "Reorganizing PG&E as a Customer-Owned, Mutual Benefit Corporation," PowerPoint presentation, October 2019.

238 **had an immediate rapport:** Richard, interview.

238 **Richard followed up with more information:** Richard, interview.

238 **On October 21, 2019:** Rebecca Smith, "San Jose to Propose Turning PG&E into Giant Customer-Owned Utility," *Wall Street Journal*, October 21, 2019.

239 **They sent a letter to Batjer:** Liccardo et al. to Batjer and commissioners, "RE: Critical Matters Related to the PG&E Bankruptcy," November 4, 2019, sanjoseca.gov/Home/Components/News/News/559/4960?npage=3&arch=1.

Twelve: A Fragile Deal

242 **On October 8, 2019:** Letter to the court, Doc. 4468, October 24, 2019, United States Bankruptcy Court, Northern District of California, Case No. 19-30088.

243 **He wondered if he couldn't:** Will Abrams, interview with the author.

243 **The hearing began promptly:** Hearing transcript, Doc. 4467, October 24, 2019, United States Bankruptcy Court, Northern District of California, Case No. 19-30088.

243 **In September, PG&E had come:** Chapter 11 Plan of Reorganization, Doc. 3841, September 9, 2019, United States Bankruptcy Court, Northern District of California, Case No. 19-30088.

243 **The plan proposed settling victims':** Peg Brickley and Katherine Blunt, "PG&E Plan Offers Nearly $18 Billion to Wildfire Victims and Public Entities," *Wall Street Journal*, September 9, 2019.

244 **Abrams was run almost single-handedly:** Rob Copeland, "Hedge-Fund World's One-Man Wealth Machine," *Wall Street Journal*, June 2, 2014.

244 **PG&E's bondholders were pushing:** Peg Brickley and Katherine Blunt, "PG&E Bondholders Ally with Wildfire Victims to Propose New Bankruptcy Exit Plan," *Wall Street Journal*, September 19, 2019.

244 **That would wipe out:** Peg Brickley and Katherine Blunt, "Alliance of Bondholders and Fire Victims Unsettles PG&E Bankruptcy," *Wall Street Journal*, September 23, 2019.

247 **So he went home:** Abrams, interview.

249 **Watts homed in on PG&E:** Katherine Blunt, "As PG&E Fire Victims Weigh Settlement, Lawyer's Role Attracts Scrutiny," *Wall Street Journal*, May 11, 2020.

250 **Trostle, a longtime Butte County:** Kirk Trostle, interview with the author.

251 **They had assembled:** This presentation was recorded and uploaded to Vimeo as a password-protected video viewed by the author.

257 **It was a formal objection:** United States Bankruptcy Court, Northern District of California, San Francisco Division, Case No. 19-30088, Dec. 16, 2019, Document No. 5139.

257 **The lawyer, wearing slip-on loafers:** Abrams, interview.

257 **Abrams waited patiently:** Objection to Debtors' Motion, Doc. 5169, December 17, 2019, United States Bankruptcy Court, Northern District of California, Case No. 19-30088.

259 **anguished letters began appearing:** Letters to the Court, Doc. 5739, February 10, 2020, United States Bankruptcy Court, Northern District of California, Case No. 19-30088.

260 **On New Year's Eve, he created:** Amended and Renewed Objection of William B. Abrams, Doc. 5533, Ex. A, January 28, 2020, United States Bankruptcy Court, Northern District of California, Case No. 19-30088.

260 **Abrams filed a lengthy objection:** Amended and Renewed Objection of William B. Abrams, Doc. 5533, January 28, 2020, United States Bankruptcy Court, Northern District of California, Case No. 19-30088.

260 **Montali invited Abrams to speak:** Transcript of Proceedings, Doc. 5634, February 5, 2020, United States Bankruptcy Court, Northern District of California, Case No. 19-30088.

263 **Frankly, Trostle didn't care:** Trostle, interview.

263 **One prominent law firm:** Skikos, Crawford, Skikos & Joseph, email to clients, May 7, 2020.

265 **Watts came out vigorously:** Blunt, "As PG&E Fire Victims Weigh Settlement."

265 **Centerbridge never revealed:** Blunt, "As PG&E Fire Victims Weigh Settlement."

Thirteen: Twice Convicted

267 **Noel put in a call:** Marc Noel, interview with the author.

267 **Noel was quick to give:** Noel, interview.

267 **But Haggarty didn't think:** James Haggarty, interview with the author.

267 **Haggarty advised Noel:** Haggarty, interview; Noel, interview.

268 **His dining room had become:** Noel, interview.

268 **Noel put out an email:** Noel, interview.

269 **Noel and Ramsey called:** Noel, interview.

269 **Noel wrapped up his interview:** Noel, interview.

270 **One day, Ramsey came across:** Mike Ramsey, interview with the author.

270 **It was from an engineer:** Email from Carlos Gonzalez to Manho Yeung et al., "EO-RIBA Check-in with Manho," Feb. 25, 2014.

271 **Within a year, the jurors:** Unless otherwise noted, details of PG&E's maintenance and inspection practices are drawn from the *Camp Fire Public Report,*
here: Butte County District Attorney, *The Camp Fire Public Report: A Summary
of the Camp Fire Investigation,* June 16, 2020, 14, buttecounty.net/Portals/30
/CFReport/PGE-THE-CAMP-FIRE-PUBLIC-REPORT.pdf? ver
=2020-06-15-190515-977. PG&E, at the request of U.S. District Judge William
Alsup, filed a paragraph-by-paragraph response to this report. PG&E disputed
or clarified some of the findings, but most often, the company asserted it did
not have enough information to confirm or deny them. I cite points the company disputed, but not those it couldn't confirm or deny.

271 **The company's inspection and patrol:** *Camp Fire Public Report,* 23.

272 **One of the hooks was:** *Camp Fire Public Report,* 24.

272 **The manufacturer's assessment:** Response by Pacific Gas and Electric Company, Doc. 1128, Ex. C, December 19, 2019, *United States of America vs. Pacific
Gas & Electric Company,* 3:14-cr-00175-WHA.

272 **The new guidelines substantially reduced:** *Camp Fire Public Report,* 24.

273 **In its search for records:** *Camp Fire Public Report,* 36.

273 **One of them, a former troubleman:** *Camp Fire Public Report,* 40.

274 **Minimizing those costs:** *Camp Fire Public Report,* 60. PG&E, in its response to
Judge Alsup, denies that it inappropriately "bundled expense budget projects
with capital budget projects in order to charge the expense budget costs to the
capital budget project," noting that "bundling projects for efficiency and to
avoid repeated service disruptions is appropriate, and it is reasonable to structure capital projects in ways that avoid or reduce expenses." See Response by
Pacific Gas and Electric Company, Doc. 1232, 26, July 1, 2020, *United States of
America vs. Pacific Gas & Electric Company,* 3:14-cr-00175-WHA.

274 **In an internal email, he:** *Camp Fire Public Report,* 34.

275 **He tried to push back:** *Camp Fire Public Report,* 34.

275 **The engineer kept reminding:** *Camp Fire Public Report,* 52.

276 **That same year, a finance manager:** *Camp Fire Public Report,* 54.

276 **company took a more proactive:** *Camp Fire Public Report,* 59.

277 **By their estimation:** *Camp Fire Public Report,* 79.

277 **Ramsey called it decision-making "by committee":** *Camp Fire Public
Report,* 80.

278 **In Ramsey's words, the hooks:** *Camp Fire Public Report,* 80.

278 **In the spring of 2020:** Ramsey, interview; Katherine Blunt, "PG&E to Plead
Guilty to Involuntary Manslaughter Charges in Deadly California Wildfire,"
Wall Street Journal, March 23, 2020.

279 **To his surprise, the attorneys:** Ramsey, interview; and Katherine Blunt,
"PG&E to Plead Guilty."

279 **seven rows of twelve thumbnails:** Ramsey displayed the same slides
during a post-hearing press conference. "Press Conference following the
PG&E Arraignment," YouTube video, 1:26:02, streamed live on June 16,
2020, posted by Butte County District Attorney, youtube.com/watch?v=Nu
Xeau92GYk.

281 sell about $3.25 billion in shares: PG&E, "PG&E Corporation Announces $3.25 Billion Common Stock Investment from Multiple Investors," *PG&E Currents*, press release, June 8, 2020.

281 At the bankruptcy hearing: United States Bankruptcy Court, Northern District of California, San Francisco Division, Case No. 19-30088, Feb. 5, 2020, Document No. 7984.

281 Will Abrams was deeply skeptical: Will Abrams, interview with the author.

285 To Noel, it seemed: Noel, interview.

286 Noel had printed out: Noel, interview; author has seen the framed copy.

Epilogue

287 The company didn't flag it: Transcript of Proceedings, Doc. 1467, September 14, 2021, *United States of America vs. Pacific Gas & Electric Company*, 3:14-cr-00175-WHA.

287 Hours later, around 11:00 a.m.: Transcript of Proceedings, Doc. 1467, Ex. X-3, September 14, 2021, *United States of America vs. Pacific Gas & Electric Company*, 3:14-cr-00175-WHA.

287 When it looked as though: Transcript of Proceedings, Doc. 1467, September 14, 2021, *United States of America vs. Pacific Gas & Electric Company*, 3:14-cr-00175-WHA.

289 The trust hadn't yet sold: Katherine Blunt, "PG&E Wildfire Victims Still Unpaid as New California Fires Weigh on Company's Stock," *Wall Street Journal*, August 11, 2021.

289 The fire, he wrote, should come: William B. Abrams Motion to Enforce Disclosure Requirements or Reconstitute the Fire Victim Trust Oversight Committee, Doc. 1467, August 2, 2021, *United States of America vs. Pacific Gas & Electric Company*, 3:14-cr-00175-WHA.

290 "We made history": Marc Noel, interview with the author.

291 Large reservoirs reached historically: Katherine Blunt and Jim Carlton, "West Risks Blackouts as Drought Reduces Hydroelectric Power," *Wall Street Journal*, June 18, 2021.

292 She set the photo: J. D. Morris, "Inside PG&E, Executives Race to Get ahead of Unending Wildfire Risk and Put Power Lines Underground," *San Francisco Chronicle*, October 10, 2021.

294 The announcement left the board: Katherine Blunt, "One of America's Toughest CEO Jobs: Fixing PG&E," *Wall Street Journal*, November 5, 2021.

295 Poppe estimated PG&E's undergrounding: Katherine Blunt, "PG&E, in Reversal, to Bury Power Lines in Fire-Prone Areas," *Wall Street Journal*, July 21, 2021.

Index